IMAGES
of America

KAISER STEEL
FONTANA

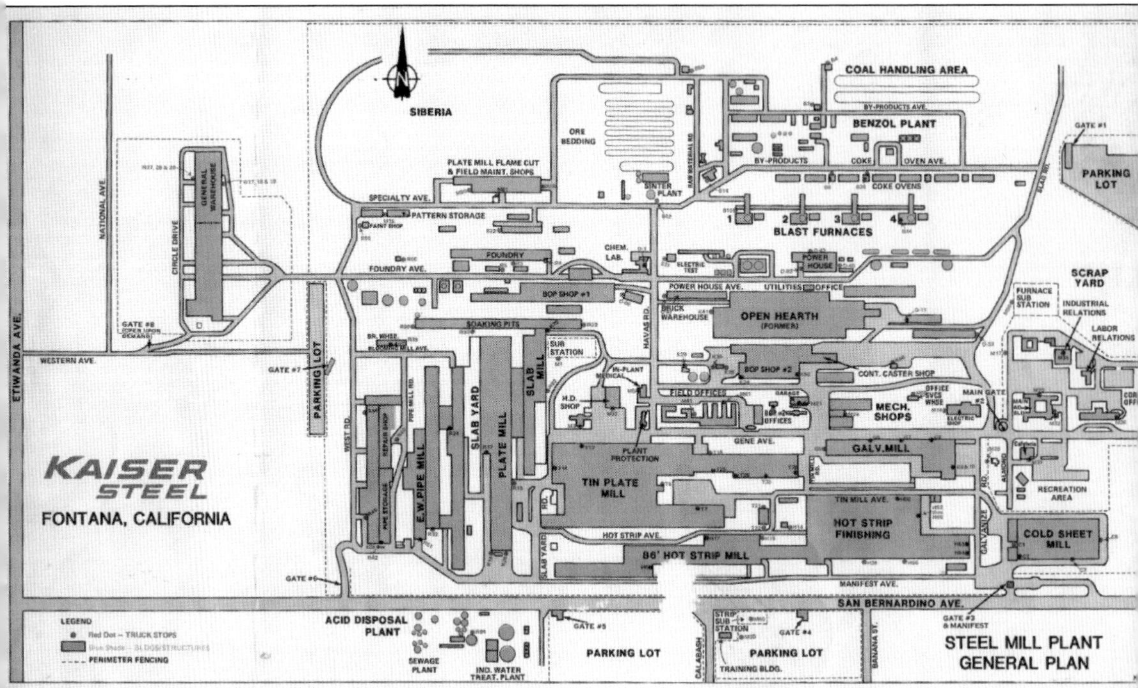

This is a *c.* 1980s map of the Kaiser facility.

TIMELINE

1941 Kaiser Company, Inc. established a base of operations in Fontana, California

1942 The official name of the Fontana operations was specified to Kaiser Company, Inc., Iron & Steel Division

1949 The official name of the Fontana mill was changed to Kaiser Steel Corporation.

1956 The official name of the overall company, with operations in many states, became Kaiser Steel Industries, Inc.

1983 After the Fontana mill closed, the Kaiser-owned operations in Fontana were renamed Kaiser Steel Resources, Ltd.

2000 Kaiser's Fontana business operations were renamed Kaiser Ventures. Kaiser Ventures continues to operate in Fontana.

ON THE COVER: The by-products area, the coke ovens, and the nearby handling breaker building are visible in this photograph taken on February 20, 1958, with the new snowfall on the San Gabriel Mountains north of the plant. (Courtesy of Jonnie Long.)

IMAGES
of America

KAISER STEEL FONTANA

John Charles Anicic Jr.

ARCADIA PUBLISHING

ISBN 978-0-7385-4650-6

Published by Arcadia Publishing
Charleston, South Carolina

Printed in the United States of America

Library of Congress Catalog Card Number: 2006923162

For all general information contact Arcadia Publishing at:
Telephone 843-853-2070
Fax 843-853-0044
E-mail sales@arcadiapublishing.com
For customer service and orders:
Toll-Free 1-888-313-2665

Visit us on the Internet at www.arcadiapublishing.com

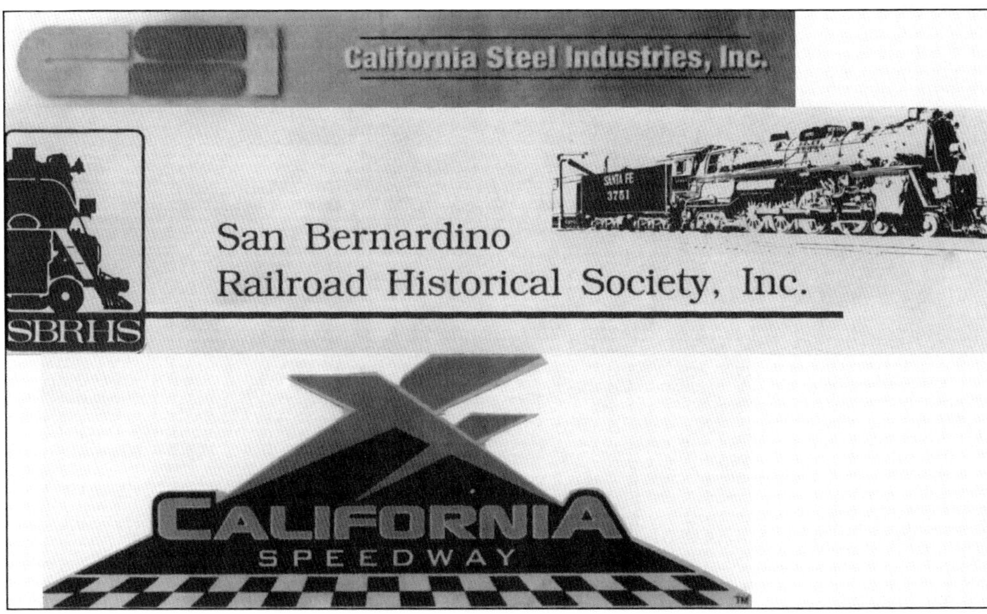

The logos of the companies and businesses that occupied the property generally known as Kaiser Steel included the Fontana Farms Company, who used the letter "F" in green on a white background; the Kaiser steel mill, using a letter "K"; and those of California Steel Industries, Inc.; San Bernardino Railroad Historical Society, Inc.; and California Speedway, shown here.

Contents

ACKNOWLEDGMENTS

Many people important to the history and preservation of Fontana's Kaiser Steel heritage have donated their time, money, and artifacts to this book They have helped me research and chronicle on these pages the company's varied history in the city of Fontana. I never worked at Kaiser, and only a cousin's husband from my family worked in the mill. But past employees have shown me great courtesy that I could not have guessed would happen. They were happy to help me and happy someone was finally doing something to preserve the mill's story. I never imagined I would do two history books as a first-generation American.

Without the people mentioned below, this book would not have existed. If I have forgotten anyone, I am sorry. This book and the contributions made by these people have helped start the Fontana Heritage Museum Association, which preserves the history of the city, Kaiser Steel, and the local rabbitry. Nearly 80 persons wanted to help, and I have yet to get back to some of them. Thanks go to the Roy Dowd family and to Judith McOmie Boyle and Cornelius DeBackey of the *Herald News*. The following persons were a great help: Judy Afersa, Ann Blazer, Bill Boatman, Lorraine Boess, Alan Bone, Edward and Lois Cook, Ruth Covert, Ann Craig, Shirley J. Croteau, Charles Dall, Jeanine Davis, Cathy Elliott, Jackie Ficarotta, Danny Flores, other members of the Flores family, Joan Gill, Trush Gilmore, Dick Horzen, Regina Hostetler, Larry Husted, Steve and Linda Ipson, Howard Herman Jordan, Alexia King, Jonnie Long, Joan McCarthy, Phyllis McKee, Mildred McKee, Norma Matousek, Julian Mendoza, Lorretta Morris, Dale Pipler, Lena Pock, Earl Ranney, Ray Robertson, Jim Rodriquez, Steven Scalf, Norma Scanlon, Joe Trebizo, William Videgar, Frank Videgar, Mary Lou Videgar, Rerha Weaver, and Margie Williamson.

Special gratitude goes to John Rapoza, a retired Santa Fe Railroad engineer; Don Black and Bob Kitell, who are associated with the 3751 Engine; Richard and Karen of Burnett's Photography; Kaiser Hospital historian Tom Dedley; Damon Nance of the Kaiser High School Library; and Dennis Bickmeier of California Speedway.

Extra-special thanks go to my wife, Rae Ellen Anicic, for putting up with a messed-up house while I was doing my research. I also must thank Charles Dall for all the time he has spent helping me identify photographs.

I recently received a phone call about the following persons. I don't have a photograph of them for the book, but I felt they had to be in it. I know people will remember them, as they were some of the long-employed workers at the Kaiser Steel Mill. Leona Welch started in 1942 as a secretary. She worked for 42 years, plus 5 more years part-time as a typist/clerk in the purchasing and warehouse department, for a total of 47 years. She would later become president of the Fontana Woman's Club. Delphine Powelson started working in 1943 as a secretary in the materials department. After a total of 37 years of service, she retired in 1980. A month later, she returned as a part-time employee. It is thought that both Welch and Powelson worked for a total of 50 years each. I have tried to tell the story of Kaiser Steel through the years and what happened on its 1,800 acres—the building and later decline of a great company. Some of the information came from Kaiser papers, newspapers, and pamphlets.

INTRODUCTION

On December 7, 1941, the village of Fontana was bathed in California's winter sunshine. Stretching into the distance were miles of vineyards, groves of citrus, poultry farms, and gardens. In the background, a snow-tipped line of mountain crests met a blue, cloudless sky. Amid the groves and gardens beyond the town's outskirts, a great herd of big and little pigs rooted and fed their way toward becoming pork, ham, and bacon. Fifty miles away, commercial ships sailed up and down the coast and across the blue Pacific Ocean.

The bombs then crashed in Hawaii, and their explosions were echoed by the roaring and hammering of America's furnaces, factories, and workshops as they produced weapons and all the accoutrements of war. Along the Pacific, coastline shipyards spawned or grew. As fast as they were built, ships were launched, splashing into the ocean with the urgent need for more. Some might think a year to build a ship was reasonable, yet the war could be lost within that time. What about a month? People considered that as slow under the circumstances. Could it be accomplished in a week? Most say it would be impossible, but the impossible was accomplished.

All of the steel for these ships launched into the Pacific Ocean, however, came from the ore deposits of Northern Minnesota and were then shipped to the furnaces of Pittsburgh, Pennsylvania, or Chicago, Illinois, and other steel centers far from the Pacific Coast. To complete the process, the steel was carried 3,000 miles on railroads already overburdened with men and other materials for war.

In and beyond the mountains that form the backdrop of Fontana lie vast deposits of iron ore and limestone—steelmaking ingredients. In the mountain slopes of nearby Utah were veins of coal used to make coke to fire the steel. But in the entire West, there were no facilities to make steel in one coordinated plant and in the enormous quantities needed to build ships. There was no time to construct ovens and furnaces or rolling mills. It was impossible to overcome the thousands of other obstacles that require two, maybe three years of work. These were the challenges made by wartime, forming the gauntlet that was thrown and taken up in southwestern San Bernardino County.

On April 6, 1942, the pigs and piglets were still foraging on the farm at Fontana, but, before the last squeal had died away, it was drowned out by the rattle and bang of a mighty construction project. Day after day, night after night, tier upon tier of furnace, oven, and stack arose. On December 30, eight months from a standing start, in place of the hog farm, a mighty, modern blast furnace roared its battle cry of defiance to the Axis powers.

This was accomplished by Henry J. Kaiser and the men who came from all over the United States, answering his call to come to Fontana and do the impossible, which they did. Why did the men and the women of the United States mobilize to fight a European war that spread to Africa and Asia? They had to.

Two conditions regarding the Kaiser Iron and Steel Plant are generally misunderstood. It is widely believed that the Kaiser steel mill was built with money provided by the U.S. government without expectation of repayment. The fact is that on March 19, 1942, Kaiser Company, Inc.

received its First Reconstruction Finance Corporation loan from the government. This was backed by private security and was repaid incrementally in the manner of any bank loan. Additional loans on the same basis followed as the plant expanded. Many people doubted he could do it, but Henry J. Kaiser repaid the loans early.

It is also generally thought that the Kaiser mill was a "war baby," born only to meet the national emergency created by World War II. It is true enough that had it not been for the war, the construction of a steel plant such as the Kaiser Mill in the West might have been delayed for many years. Yet it was inevitable that a steel mill of Kaiser's magnitude would be built in the West to supply expanding peacetime industries; the war just hurried the process.

Population growth created the world's fastest growing residential and commercial construction markets in the West. The population in the region grew from 2 million in 1880 to 14 million in 1940. The major growth state was California, and its engine was the Los Angeles area that accumulated more than three million new people over half a century. The Kaiser steel mill was placed at the strategic distribution point of this market and close to raw materials to save on transportation.

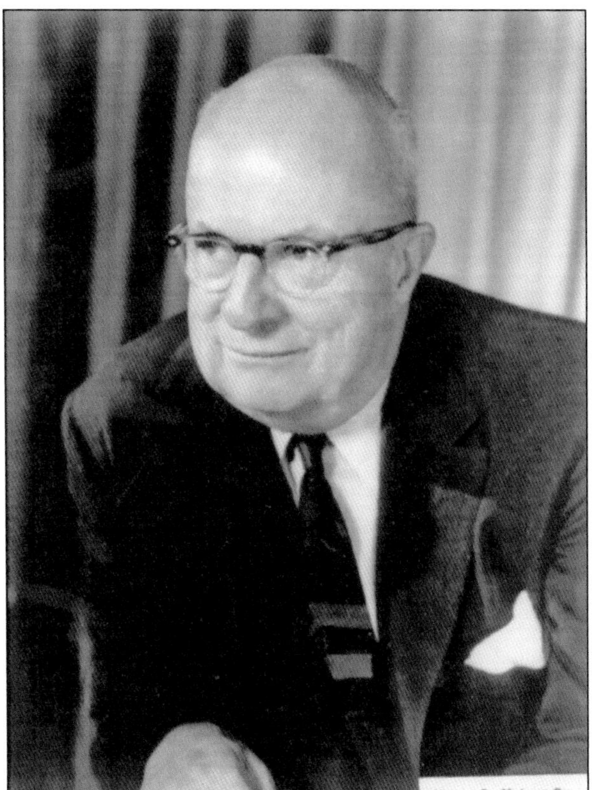

Henry J. Kaiser was born on May 9, 1882, in a white frame farmhouse at Sprout Brook, New York, one of four children of Francis J. Kaiser, a shoemaker, and Mary Yops Kaiser, a practical nurse, both immigrants from Germany. The 85-year-old industrialist died of circulatory problems on August 24, 1967, at 11:42 a.m. in Honolulu. At his side were his second wife, Alyce, and son Edgar.

One

Fontana Farms Company and Wade Hog-Breeding Farm

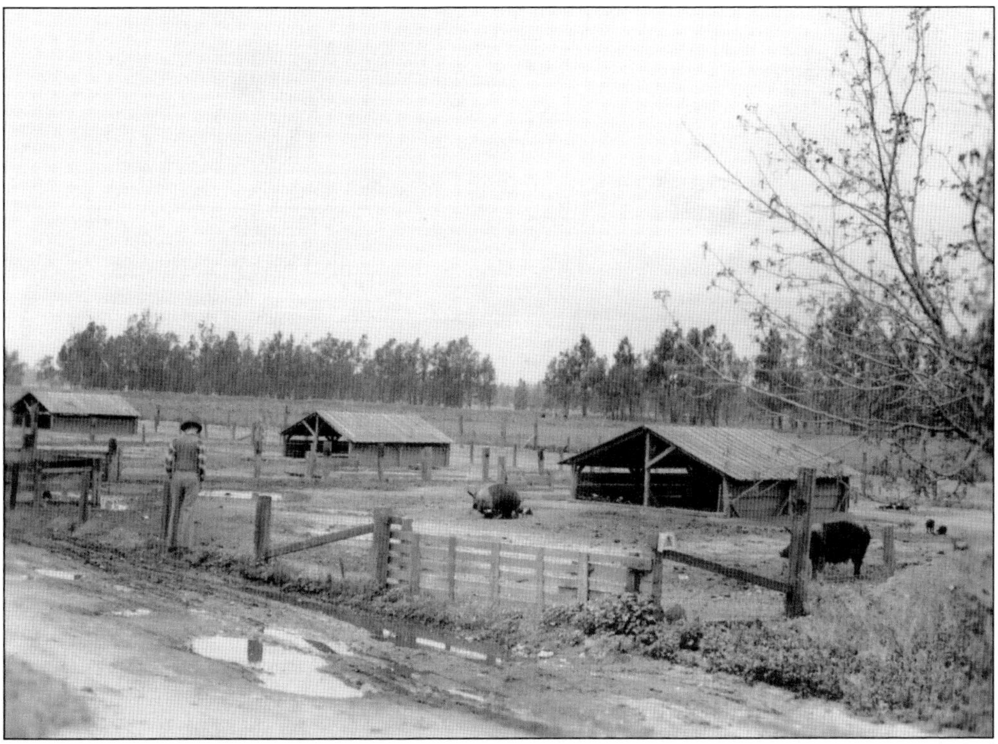

This view of the Wade Camp Hog Breeding Plant was taken on April 5, 1942. Two secluded districts existed on Fontana Farms Company land: one at the railroad siding of Declez (now South Fontana) and the other at the Wade Camp site, four miles west of the town site of Fontana. They were situated in a "neutral area," a wide zone of land comprising 1,000 acres of field crops. No land was sold to settlers, and as many as 50,000 hogs roamed the area from 1916 through the 1950s. Usually at least 30,000 pigs congregated at the Declez plants and 7,000 at Wade Camp. About 7,000 also were kept at Diamond Bar, near Pomona. (Courtesy of Alexia King.)

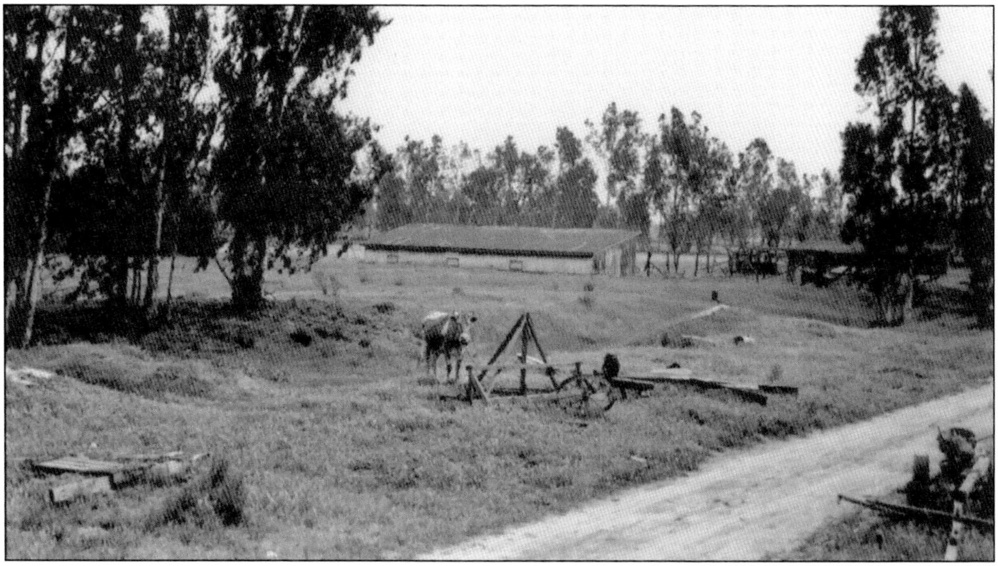

The Wade Camp was located four miles west of the town site of Fontana, on Randall Avenue west of Calabash Avenue. The 11 cabins for the workers and wives in 1941 were inhabited by South Fontana superintendent of shops Harry A. and Viola Wade; foreman Earl and Jenny Young; Zacarias and Bessie Aleman (cabin 101); Antonio L. and Rufugia Alaniz (cabin 102); Jesus and Sarah Cruz (cabin 103); Angela Silva and her son, Ralph (cabin 104); Jose Luis and Lucy G. Leyva (cabin 105); Jesus and Concha Payan (cabin 106); Angel and Mary Payan (cabin 107); Pedro Cadena (cabin 108); Parfineo and Balba Valasquez (cabin 109); Moses and Lupe Silva (cabin 110); and Cosme and Guadalupe Urguizu (cabin 111).

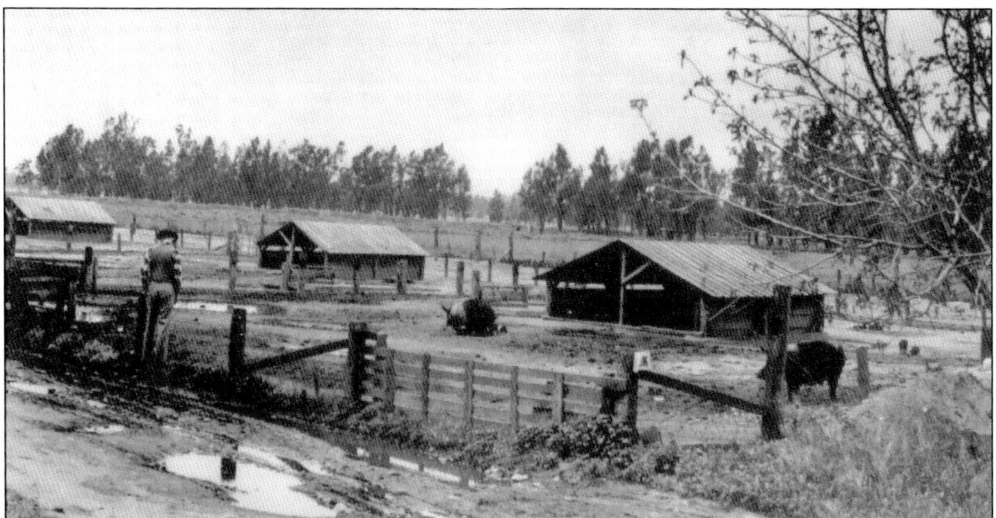

This is the Wade Hog Farm just before construction started on the steel mill in April 1942. The Wade plant was the breeding and farrowing area. Usually 100 boars and 3,000 brood sows were in the pens. As soon as the piglets were born, their numbers were recorded. At the age of 10 weeks, the piglets were shifted into the weaning unit, where they were learned to survive without their mothers. As soon as they reached approximate weights of 75 to 125 pounds, they were taken to the feeding area at Declez. More than 160 men cared the pigs at Wade and Declez. (Courtesy of Kaiser High School Library.)

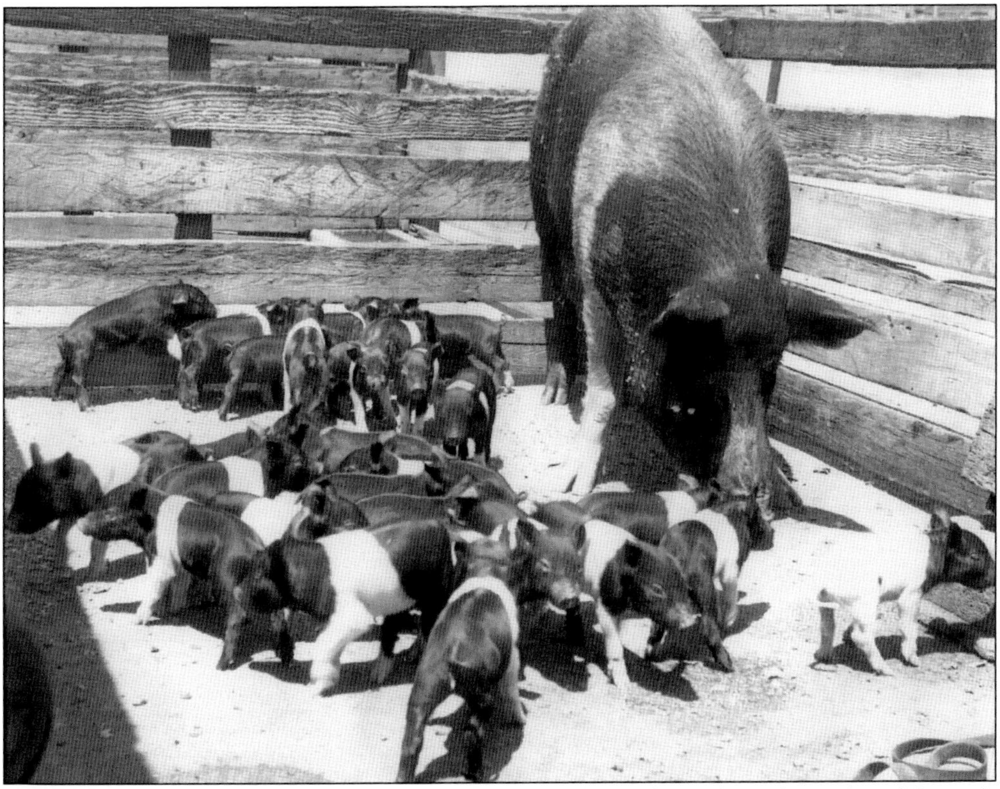

When born, the piglets were entered into a record book after being logged on a board for each pen. If one sow farrowed too many and another to few, the litters were equalized. (Courtesy of Judith McOmie Boyle.)

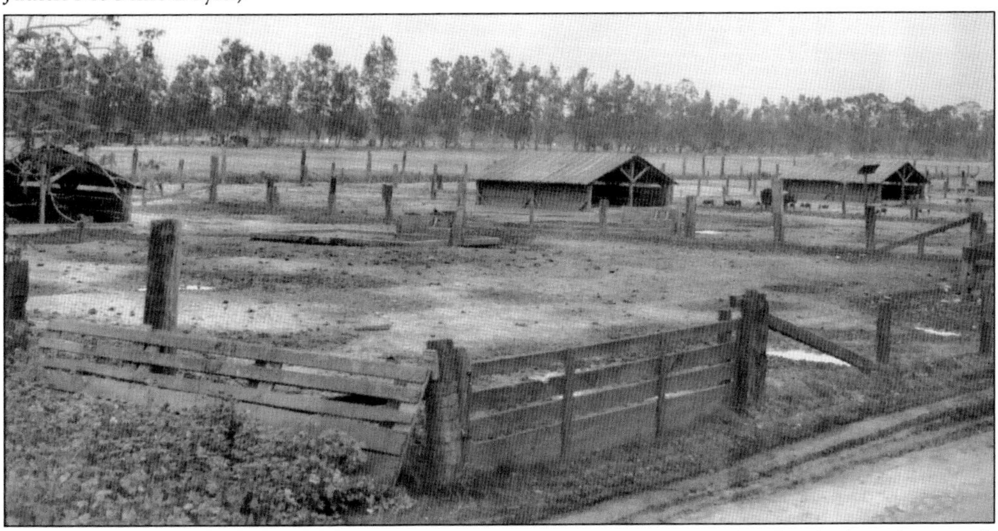

This is the Wade Hog Farm just before construction started on the steel mill. The breeding stock are direct descendents of some of the famous prize winners once owned by the Diamond Bar Ranch in the days when that holding boasted some of the porcine aristocrats of America. Most of the Fontana hogs were Durco-Jerseys, with some Poland Chinas, Berkshires, and Chesters. (Courtesy of Judith McOmie Boyle.)

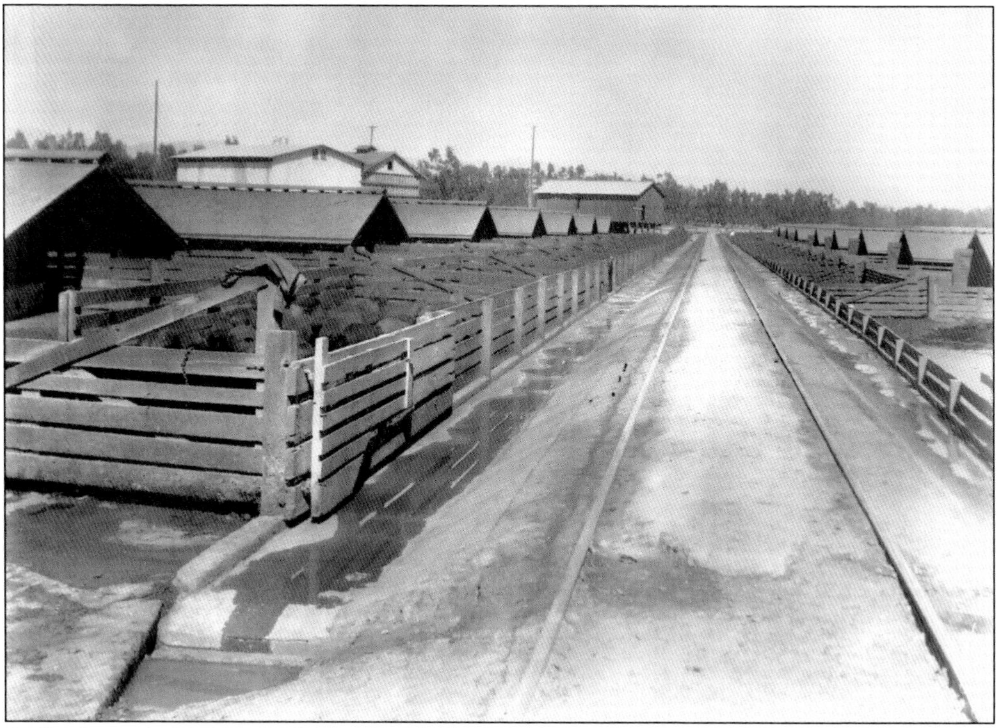

At Declez, a deep concrete ditch carried all liquid to a septic tank. The rest of the manure was shoveled out into the alleyways and a giant machine built on the grounds scraped it up and dumped it into trucks by means of scrapers on an endless chain. The trucks backed off, carrying their loads to a vast, all-concrete compost pit that was 500 feet long, 75 feet wide, and 12 feet deep. Here the material drained until only about 50 percent of the moisture remained. This partially dried manure was covered with a coating of gypsum to kill odor and insect eggs. Later it was spread on a great concrete platform where it became so dry it was sacked for fertilizer. The pens are pictured above ready for cleanup. A cleanup machine is pictured below in action. (Both photographs courtesy of Judith McOmie Boyle.)

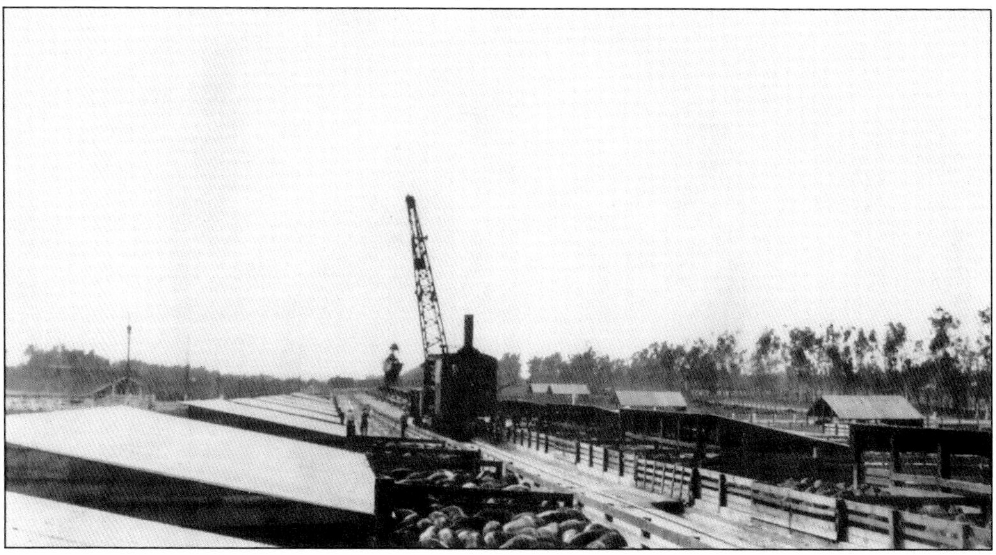

This is a close-up of the machine that cleans the pen areas. The pen floors, feeding platforms, and other surfaces, like the train gondolas, were treated to a daily bath of live steam, which effectually removed any germs that may have gained a precarious hold around the pens and sheds. Every week, the wooden posts and sides were painted with disinfectant. (Courtesy of Judith McOmie Boyle.)

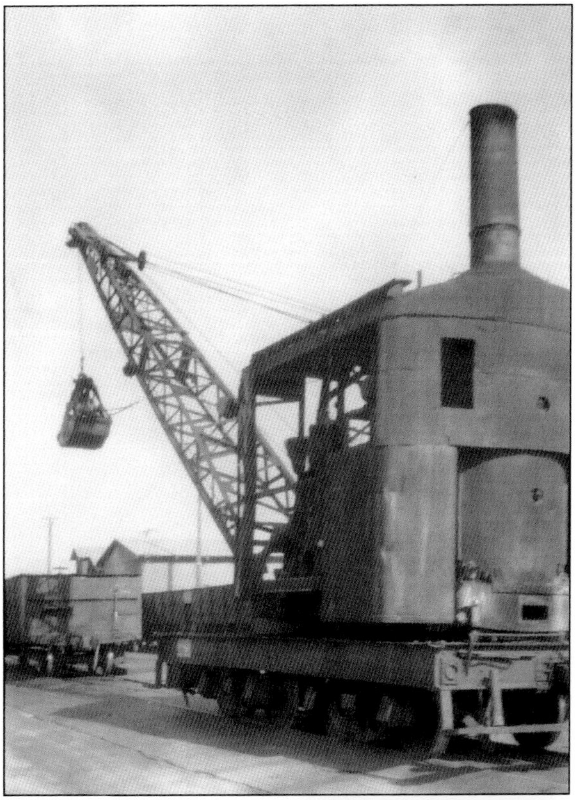

The gondolas are being cleaned here by steam power. The railroad tracks ran right through the center of the yard, and concrete filled all the spaces between the rails. Any microbe that found a home would have to be equipped with a cold chisel and sledge hammer even if it succeeded in surviving the regular steaming. The gondolas from Los Angeles, via Southern Pacific trains, arrived at 9:00 p.m. each night. (Courtesy of Judith McOmie Boyle.)

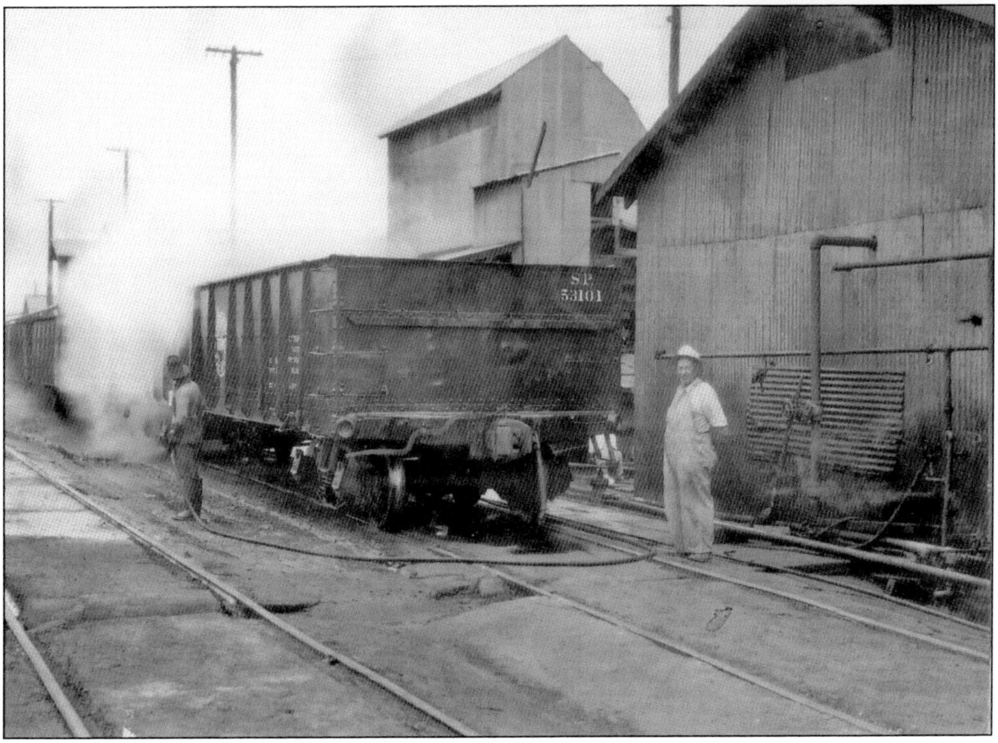

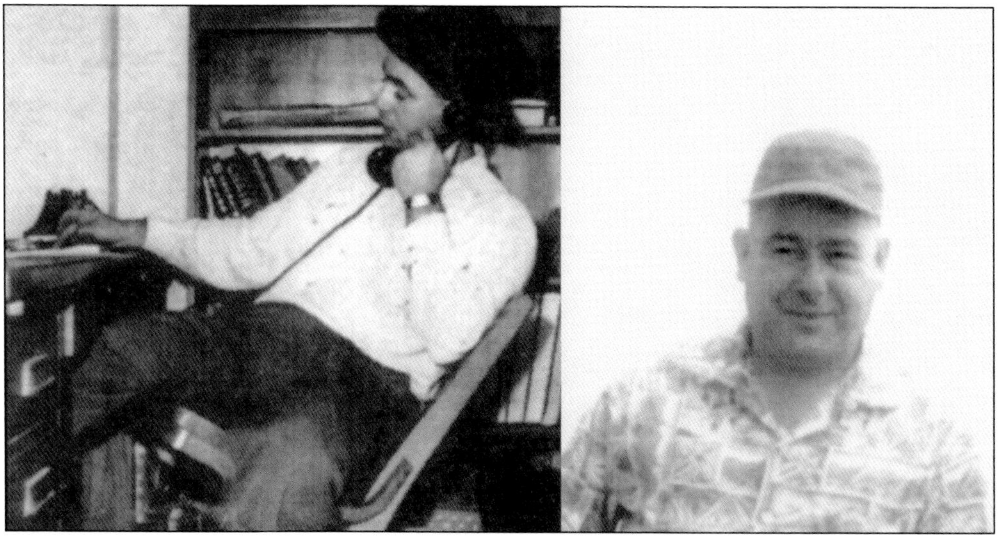

As the superintendent in charge of the warehouse and transportation, Sam Costa was a mainstay of Kaiser Steel from the beginning. Practically all building materials went through his hands to build Kaiser. He worked for Kaiser on and off from 1942 to 1955, including two other Kaiser projects. His hobbies included poker and teaching softball, and he and his wife lived in Fontana.

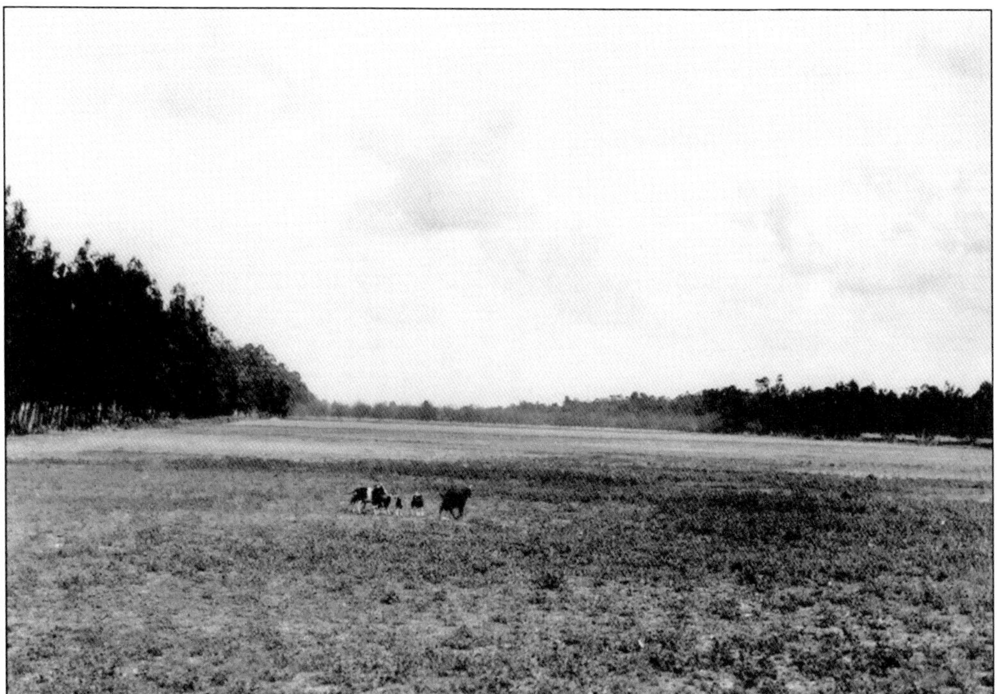

The Fontana Farms Company's land was formerly a grain-growing area. (Courtesy of Kaiser High School Library.)

Two

THE PLANT'S BEGINNING AND THE DEMISE OF THE HOG FARM

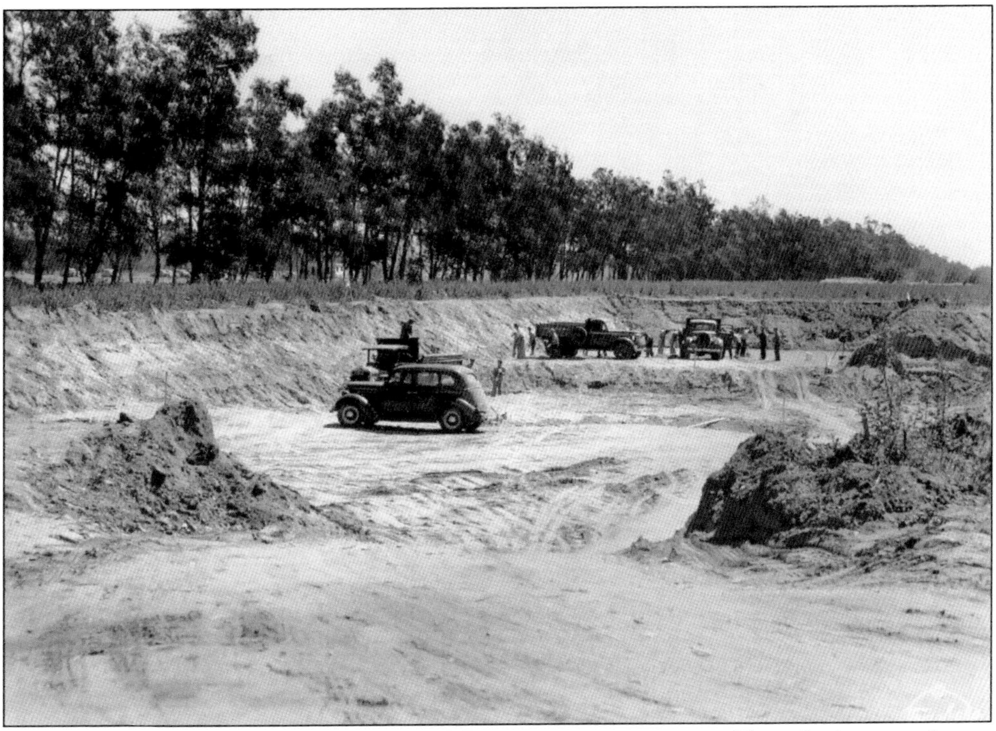

The start of excavation to ready the ground for construction is pictured here, looking northeast, on April 29, 1942. The site would accommodate the first blast furnace on the West Coast. (Courtesy of Alexia King.)

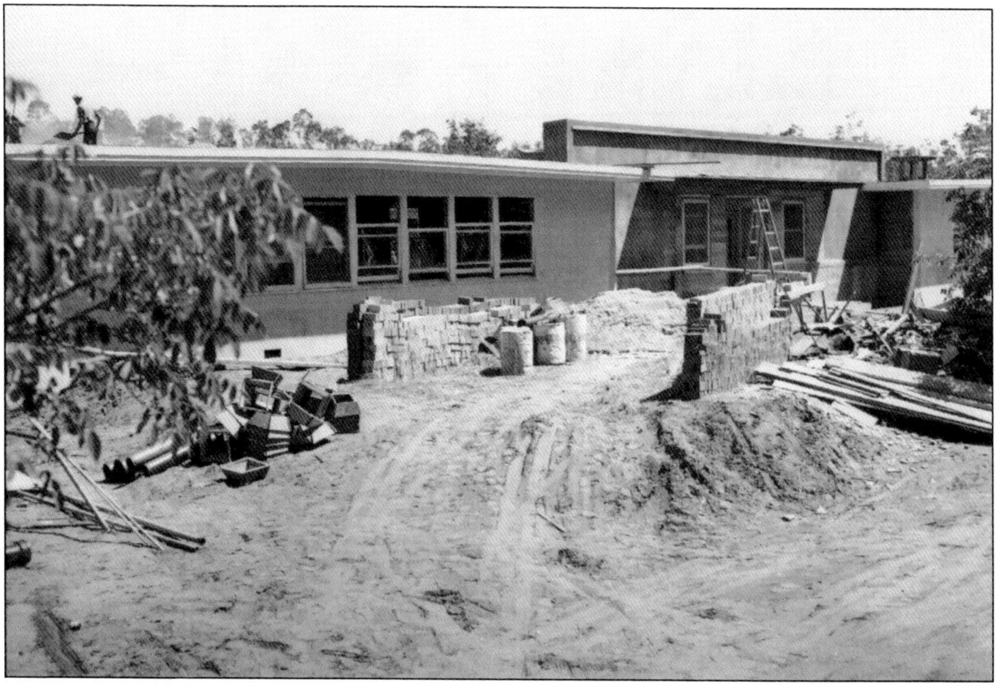

Construction of the temporary administration building and field office took place on April 29, 1942. This view looks northeast. (Courtesy of Alexia King.)

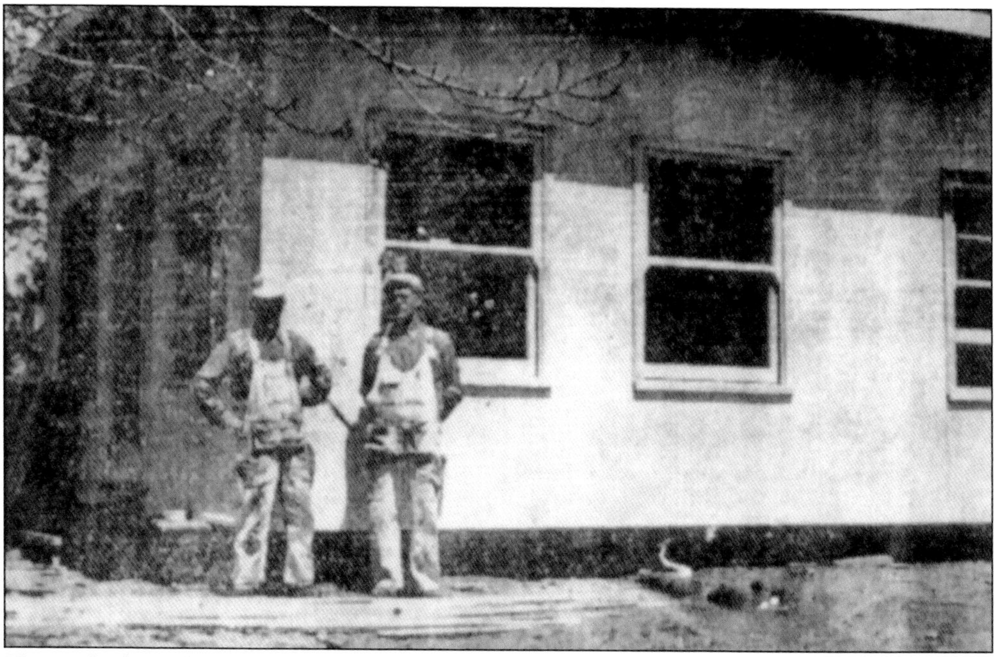

This is the first of several office buildings under construction during the preliminaries to regular construction of the $40 million steel mill. The office was built by Roy E. Dowd, a Fontana contractor. Two of Dowd's workers are pictured standing by the speedily erected structure. It was the first building constructed for Kaiser on the site and the only existing building when ground was broken for the mill. (*Daily Report* photograph; courtesy of the Dowd family.)

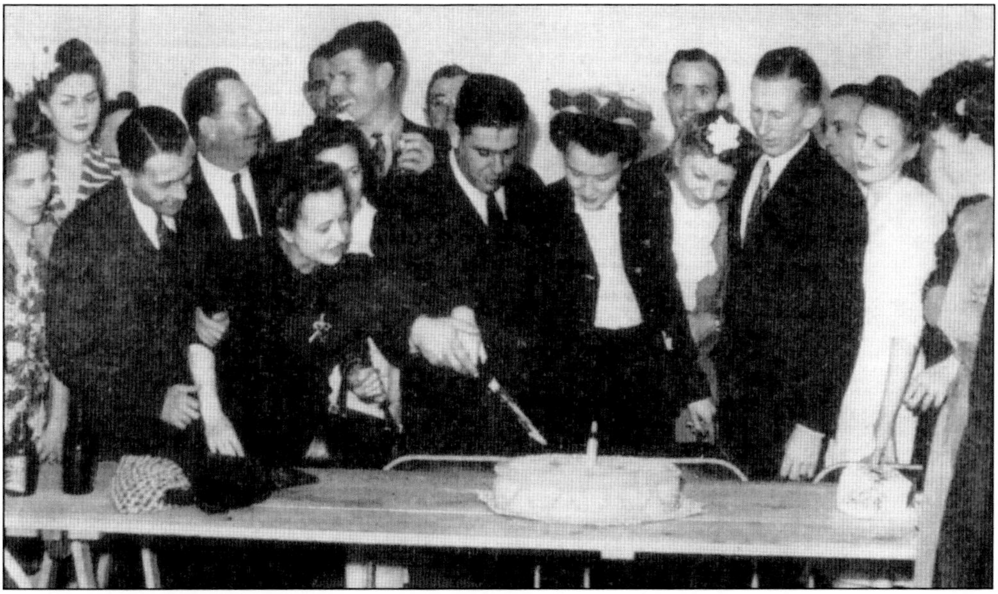

The first collective birthday for warehouse employees was celebrated at a Saturday-night dance at Lucy and John's, near Upland. Sam Costa and his wife are cutting the cake. More than 100 guests attended the celebration. The 14-inch round cake was decorated in red, white, and blue with the words, "Warehouse First Anniversary, April 9, 1943."

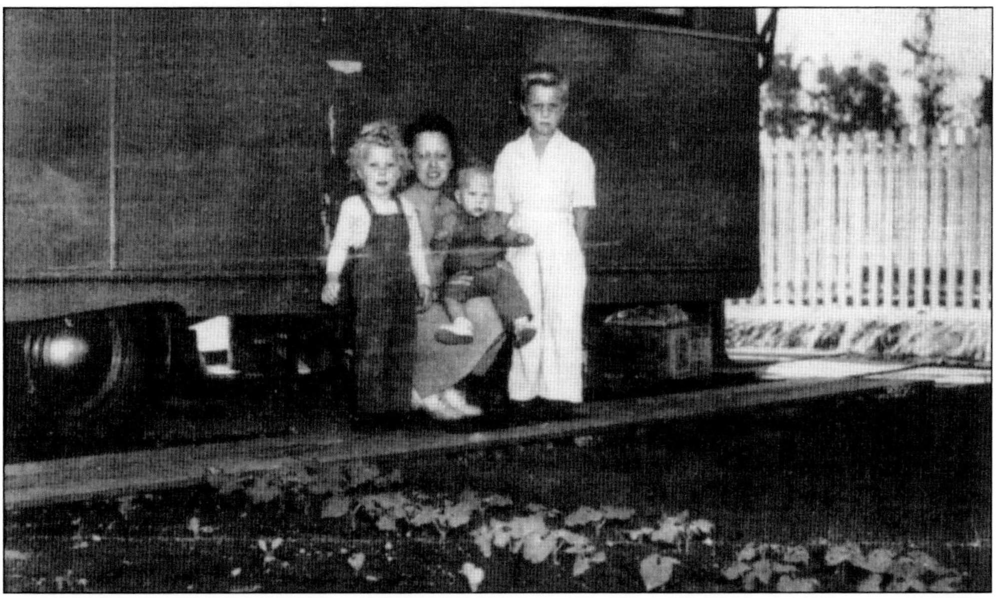

This is one of the homey trailers of "Kaiserville," the unofficial name for the area centered at Merrill and Fontana Avenues in Fontana. Carl Hoag was mayor of the unofficial community. Kaiser brought in the trailers so workers from all over the United States had a place to live. The Stamplers came to California from Akron, Ohio. Mr. Stampler was employed at the rolling mill as soon as construction was finished on it. His family, pictured here from left to right, are son Tim, three years old; Mrs. Stampler; son Ned, one year old; and son Billy, five years old. The vegetable garden at the front of the photograph was a victory garden of radishes, lettuce, spinach, beans, carrots, beets, and peas.

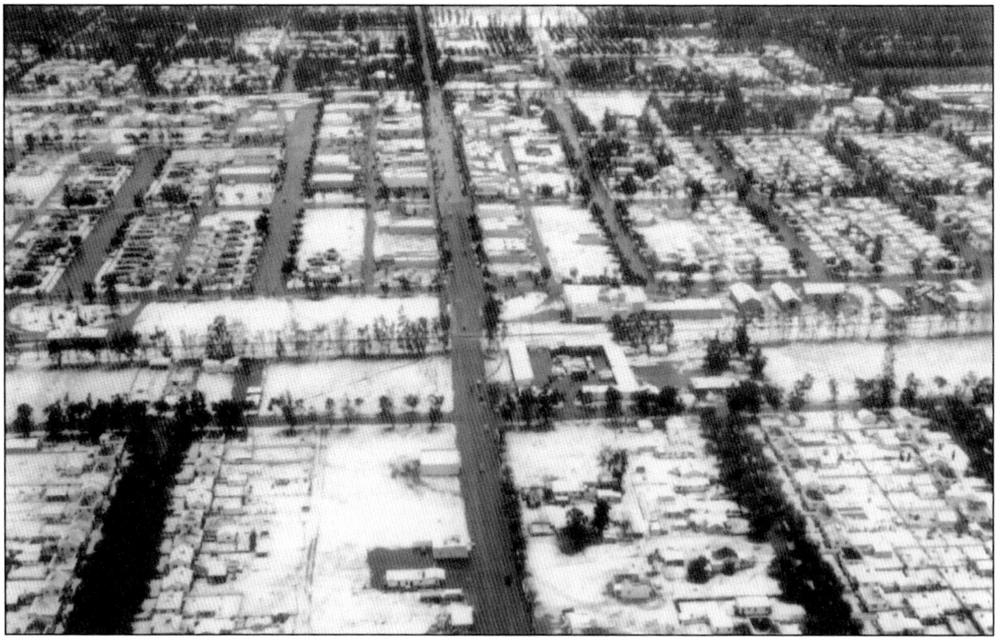

This aerial photograph of downtown Fontana, looking north, shows the city after a snowstorm around 1949. The Santa Fe Railroad tracks run horizontally through the image, while Magnolia Avenue (now Sierra Avenue) runs vertically. To the west, left of Sierra and above the tracks, are apartments. To the east, right of Sierra, is a residential area of single-family homes. All of the apartments and residences were built with federal permission and war materials. Henry Kaiser, founder of Kaiser Steel, required residences for his workers to live in. (Courtesy of Jackie Ficarotta.)

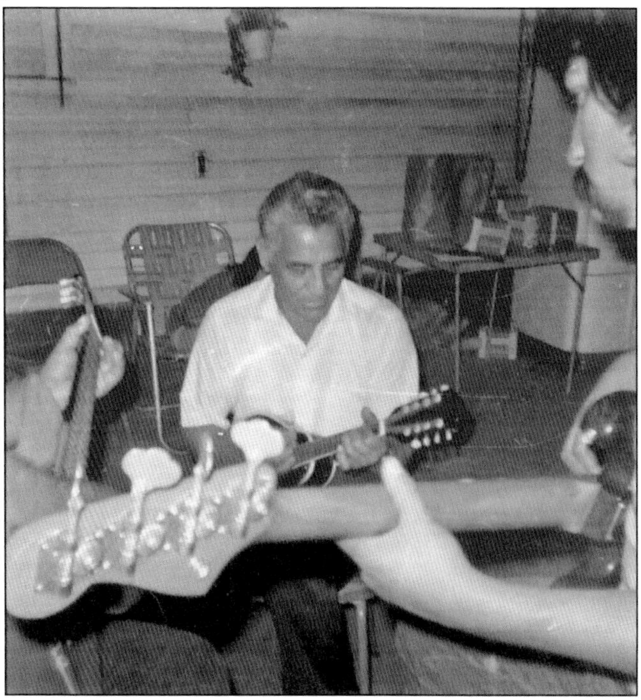

Gonsalo Flores was the foreman of the yards and roads for Kaiser. He was also in charge of constructing the first railroad tracks onto the new Kaiser property where none had previously existed. Without these tracks, no building materials could be brought to the site. Flores worked for Kaiser for 37 years until 1979. He and Sam Costa were crucial to the success of the Kaiser Mill construction. He also maintained the tracks he laid for many years. (Courtesy of the Flores family.)

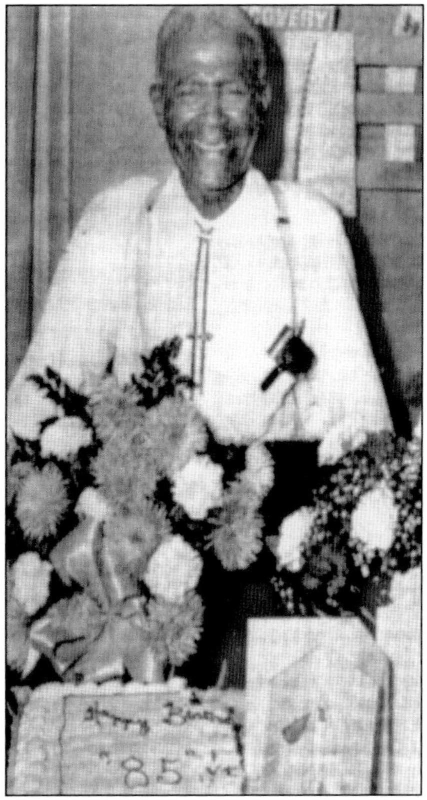

Steelman Henry J. Kaiser's first employee celebrated his 85th birthday in Fontana. James Andrew (Jimmy) Shaw received a huge birthday cake, gifts, cards, and a personal greeting from A. B. Ordway, the senior vice president and director of Kaiser Center, Inc. Shaw also worked at Shakie's Barber Shop, on 8606 Wheeler Avenue, for 14 years. His boss, Elmo L. Schreckengast, provided the cake. The special teletype from Ordway was addressed to James "Totem" Shaw. (Courtesy of Ron Brower of Don's Trim-Rite Barber.)

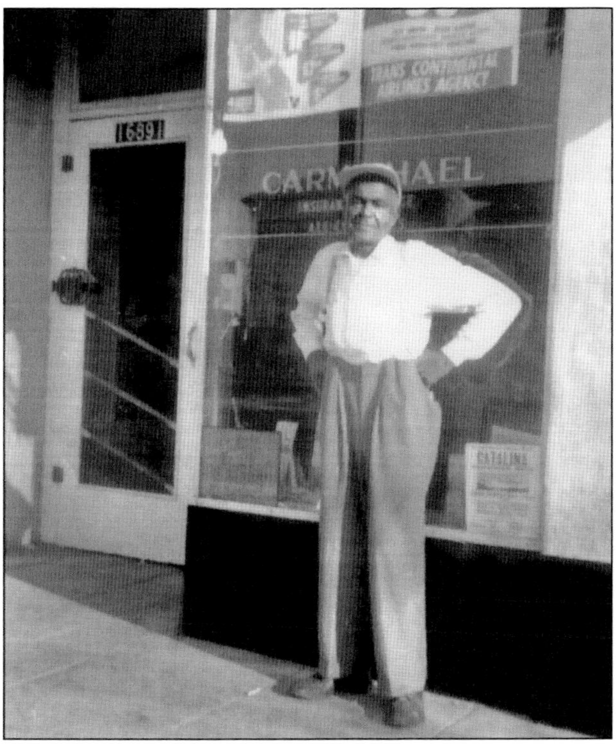

Shaw says he earned the name "Totem" when Edgar Kaiser was a small boy. "I used to carry him on my back," Shaw said. "He couldn't say, 'Pick me up,' but could say, 'Tote, tote,' which later became 'Totem.'" Shaw was put on Kaiser's payroll in 1910 as a laborer. Six were hired, but the rest quit after a week because Kaiser couldn't pay them. Shaw was with Kaiser for 40 years as a private employee and traveled everywhere with the steelmaker. "I cleaned his house, took care of the kids—any kind of work," Shaw recalled. "I spent 21 years in Illinois, then moved to Spokane, Washington, where I began work with Kaiser." He came to Fontana in 1946. He had no relatives or family. His future plans when this photograph was taken were to work another 40 years at Shakie's and then retire.

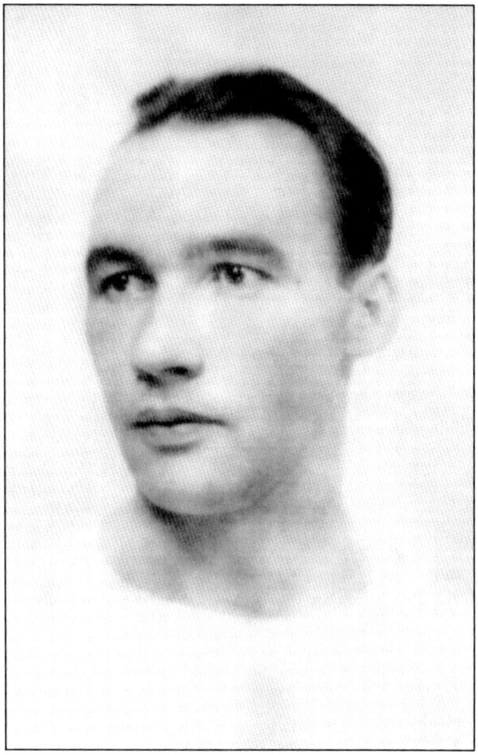

Emil Javic, who worked at Kaiser for 30 years as a chemist, is pictured here in 1943. He is the father of Jackie Ficarotta, who is the granddaughter of Stevo "Pop" Momeilovich. (Courtesy of Jackie Ficarotta.)

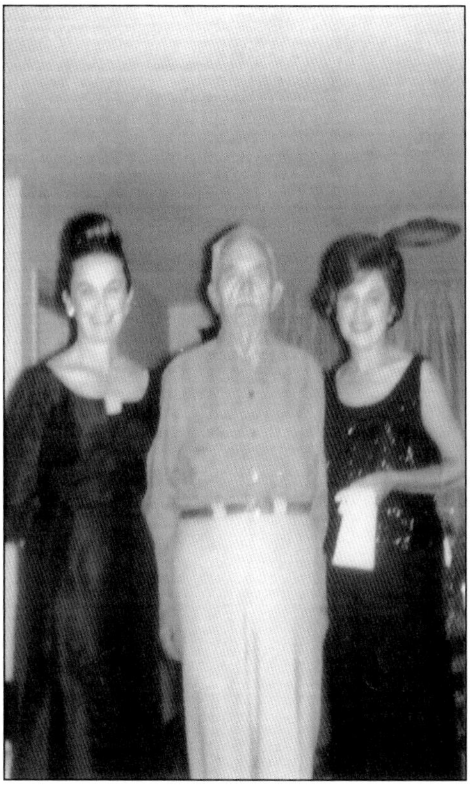

Stevo "Pop" Momeilovich and his two daughters, Mildred "Millie" Momeilovich (left) and Amelia Wassner, are pictured here. Both Mildred and Amelia worked as secretaries, the latter at the shell plant. The family used to live on Valley Boulevard below Kaiser in a large home. Many well-kept homes lined the valley in the early days of Kaiser. Stevo had Henry J. Kaiser over for dinner, who came with a special guest, singer and actor Rudy Vallee.

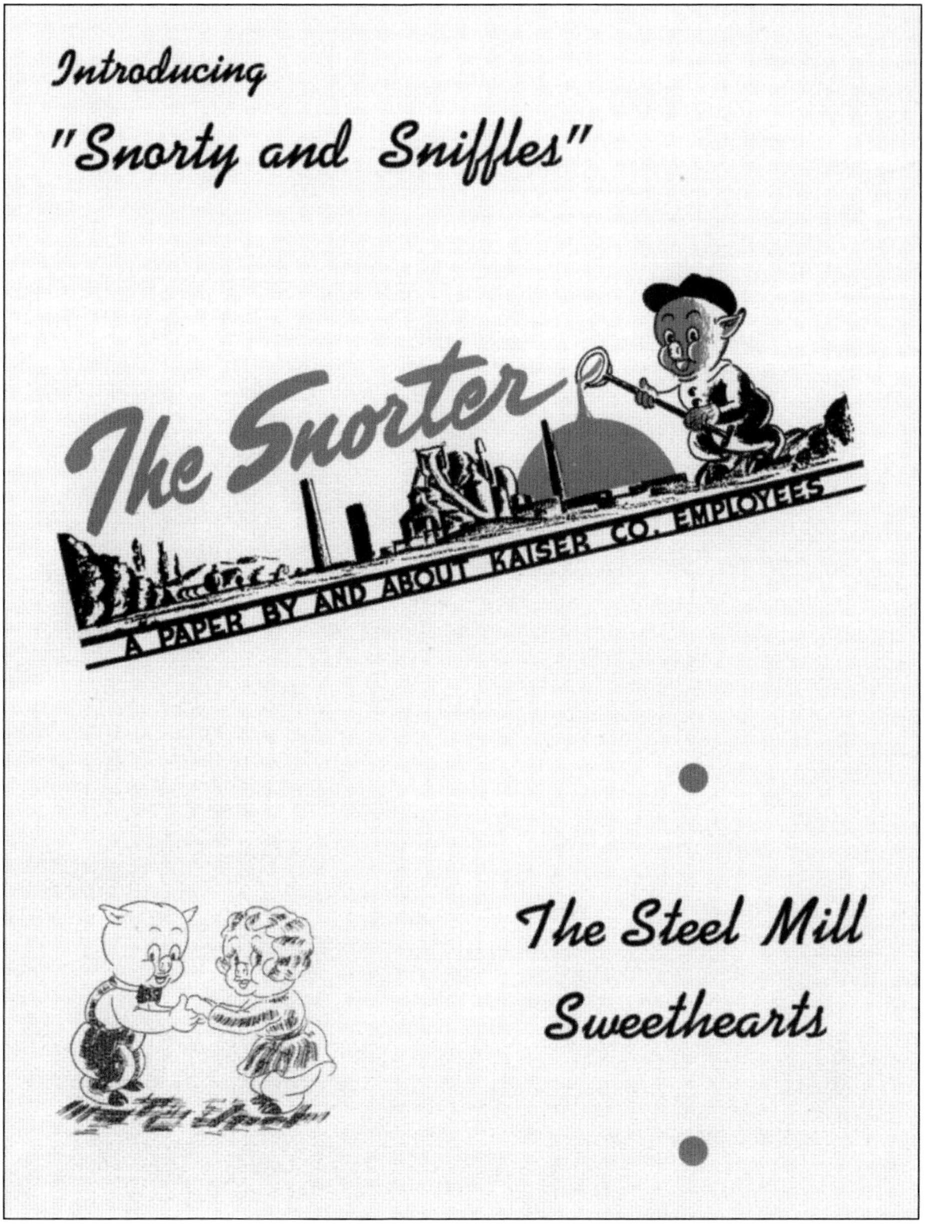

There once was a lovable little character known as Snorty, the Kaiser pig, who almost became a trademark for the steel mill. Snorty and his little sweetheart, Sniffles, stand as symbols of all the employees working in the Kaiser Iron and Steel Division. Snorty was born in December 1942 at the same time the Kaiser employee newspaper was being organized. Hundreds of names were offered, but none seemed to fit until the "Snort Valve" was suggested, and from this evolved *The Snorter.* The snort valve is a valve on a blast furnace that is opened at intervals to relieve the blast pressure when casting. Quickly grasping the idea that the blast furnace would make iron pigs, and that a live pig is sometimes known as a snorter, it did not take long for Ronnie Trombley, superintendent of the paint department, to develop a rendition of Snorty, the Kaiser pig, who has represented the Kaiser employees from that day forward. The photograph depicts a page from the booklet introducing *The Snorter.* (Courtesy of Dick Horzen.)

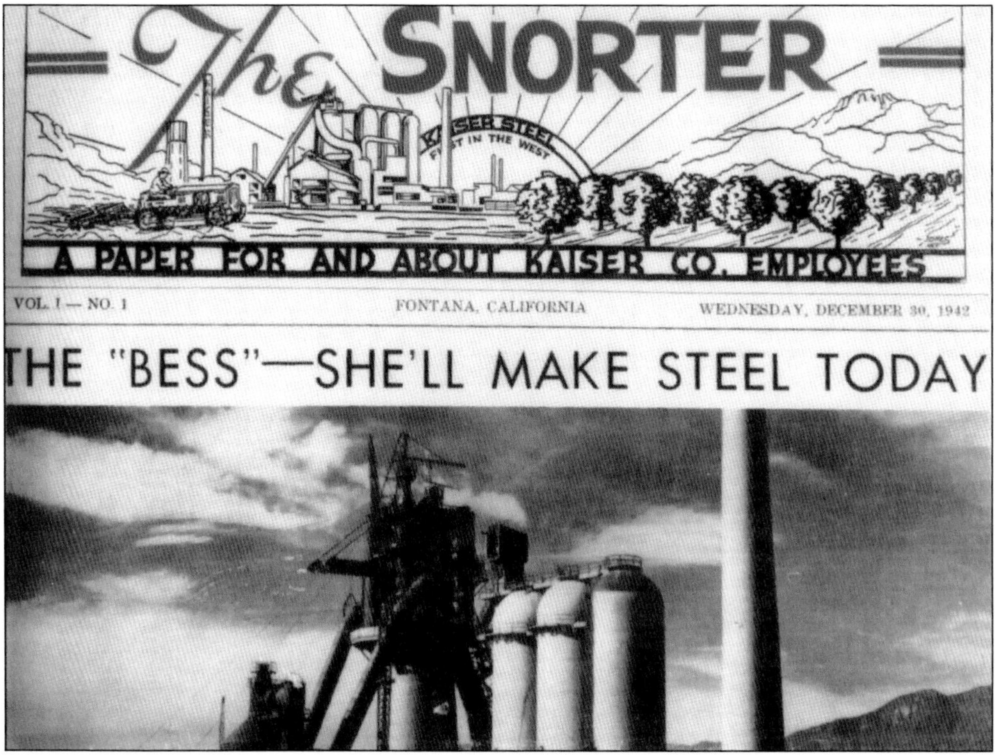

The SNORTER

KAISER STEEL FIRST IN THE WEST

A PAPER FOR AND ABOUT KAISER CO. EMPLOYEES

VOL. 1 — NO. 1 FONTANA, CALIFORNIA WEDNESDAY, DECEMBER 30, 1942

THE "BESS"—SHE'LL MAKE STEEL TODAY

This is the first-edition cover of the employee newspaper, *The Snorter*, which debuted on December 30, 1942.

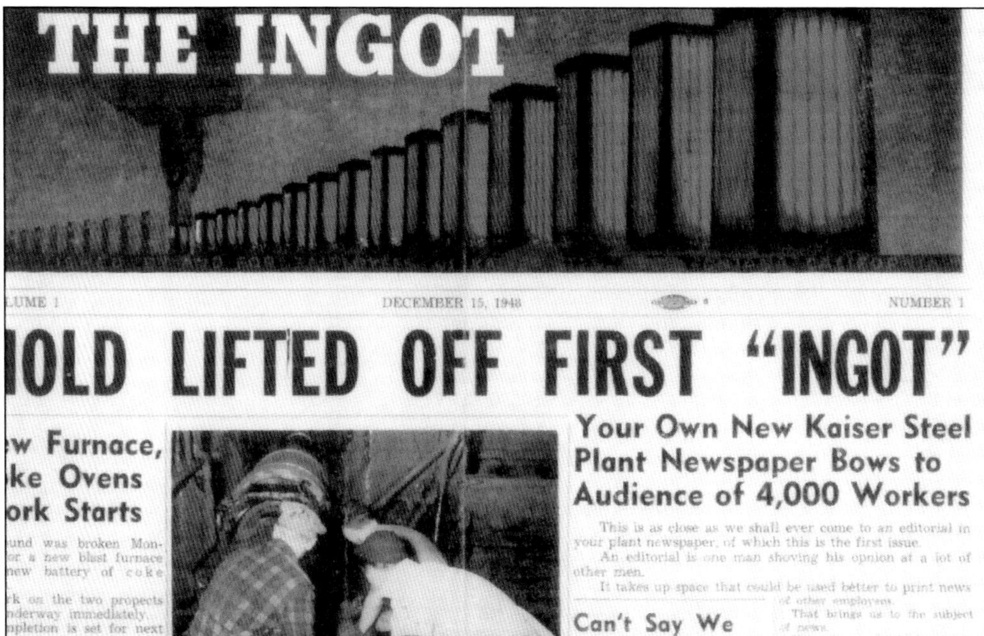

THE INGOT

UME 1 DECEMBER 15, 1948 NUMBER 1

IOLD LIFTED OFF FIRST "INGOT"

w Furnace, ke Ovens ork Starts

und was broken Mon- or a new blast furnace new battery of coke

k on the two projects nderway immediately. npletion is set for next

site is east of the pres-

Your Own New Kaiser Steel Plant Newspaper Bows to Audience of 4,000 Workers

This is as close as we shall ever come to an editorial in your plant newspaper, of which this is the first issue.

An editorial is one man shoving his opinion at a lot of other men.

It takes up space that could be used better to print news of other employees.

That brings us to the subject of news.

News is—or should be—inter- esting

Can't Say We Ganged Up on

The first edition of the employee newspaper known as *The Ingot* was distributed on December 15, 1948. The Kaiser Museum now has the printing plate for *The Ingot* in its collection.

Three

ELEMENTS OF THE KAISER MILL

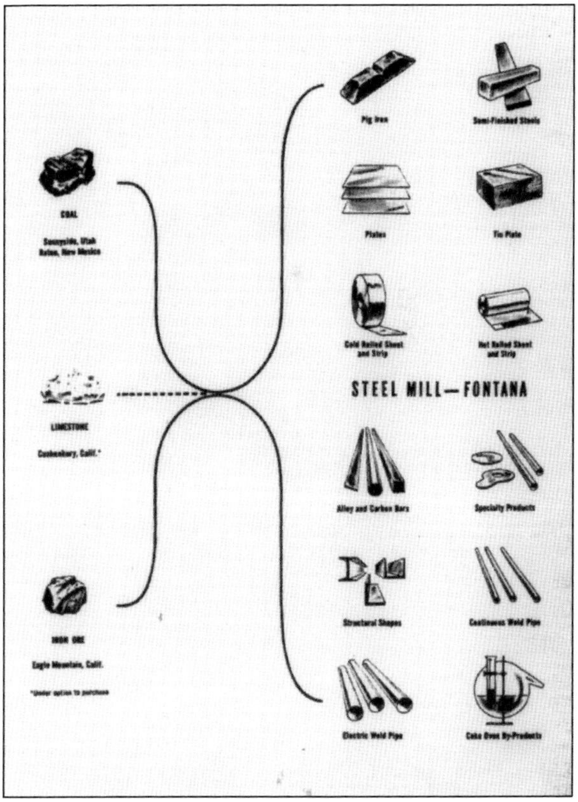

This "material chart" shows the raw materials and what they become after being combined in the steel mill's different sections. They came from deposits of iron ore, limestone, coal, dolomite, and manganese. Out of the rocks and hills of the undeveloped West came the basic ingredients for building a new industrial West.

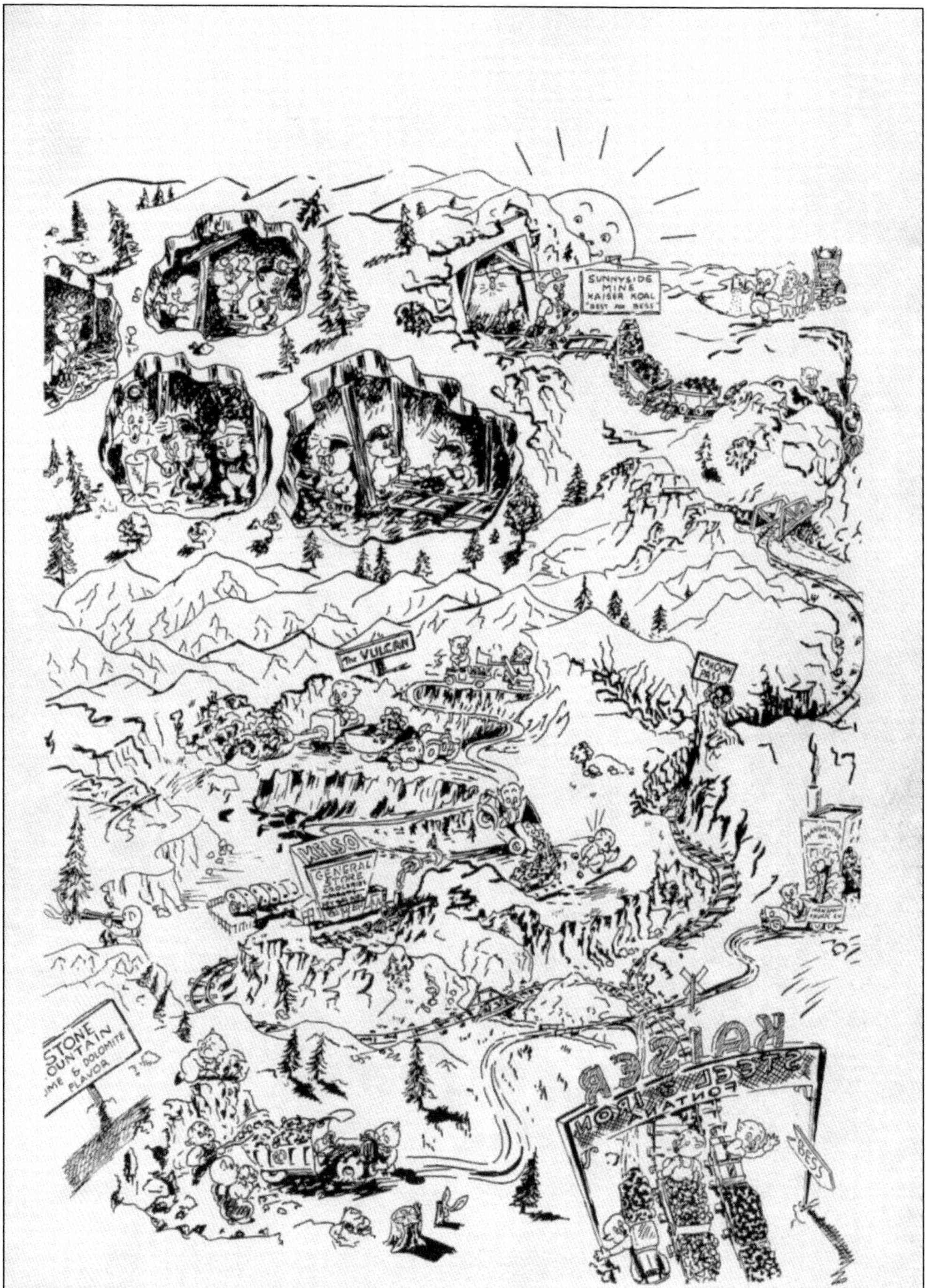

This cartoon depicts the sources of raw materials used by the mill, with the work being done by piglets.

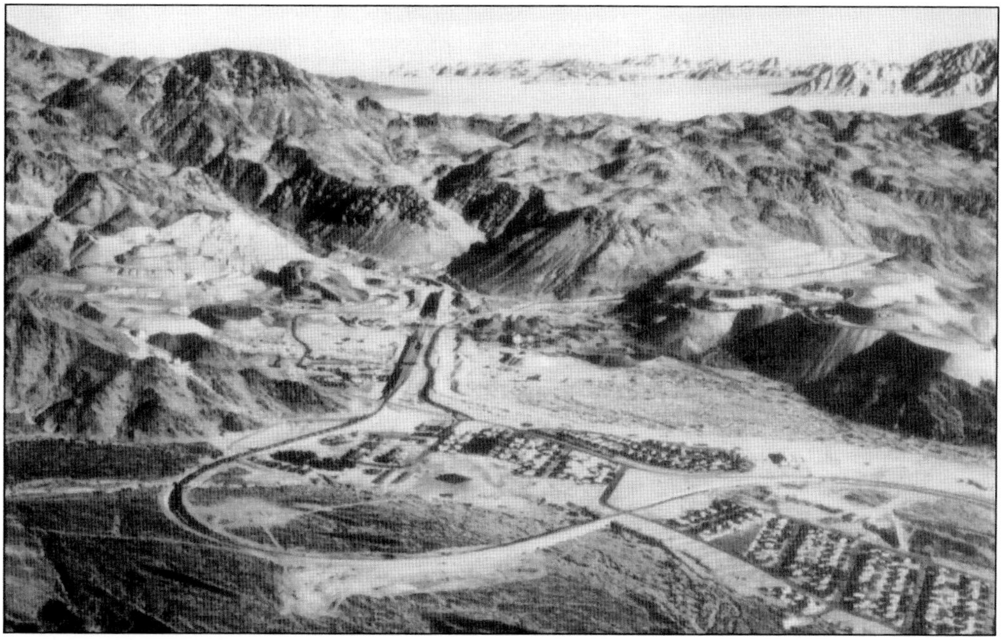

During late 19th century, prospectors' picks turned up iron ore on Eagle Mountain, near Desert Center in Riverside County. But the miners were seeking gold and had no use for what is known today as the largest deposit of iron ore in the West. The ore, which was comparable in quality to the best in the United States, lay locked in the mountains until 1944 when Kaiser Steel purchased Eagle Mountain. This aerial photograph depicts the efficient open-pit mining method at Eagle Mountain, in which overlaying rock is stripped off, exposing the ore. Holes were drilled into the rock and dynamite inserted, blasting thousands of tons at a time.

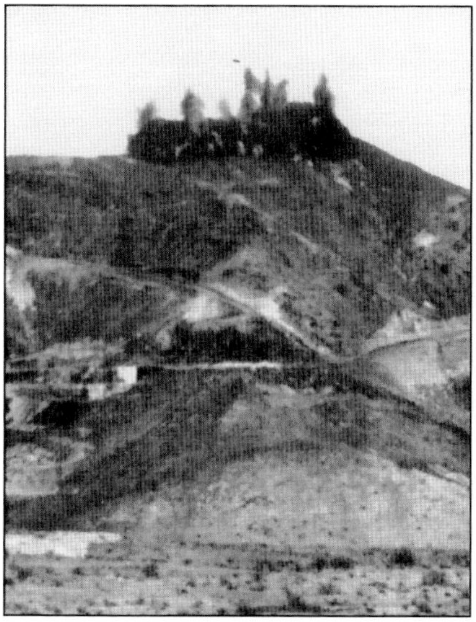

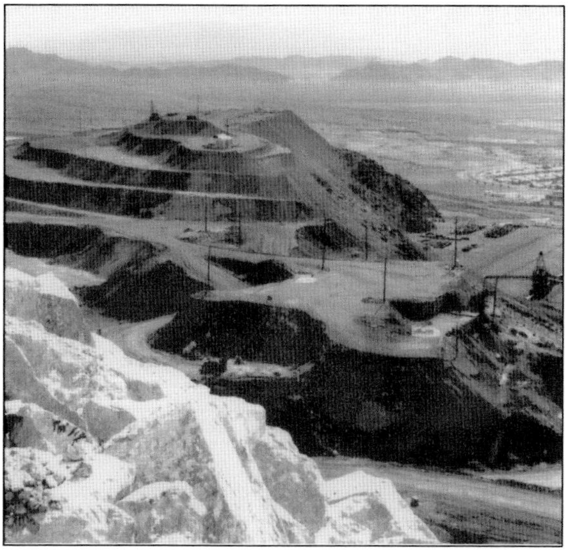

The top of the north deposit was blown off Eagle Mountain (left) in 1952, and some 30,000 tons of rock went skyward. A lot more dynamite and powder went into that mountain to expose ore in the coming years (right). Estimates of tonnage in this ore body and in the neighboring south deposits were enough to keep three blast furnaces going for 25 or 30 years. The amount of explosives used at the mines during one year cost $245,828. One blast carried the price tag of $4,000.

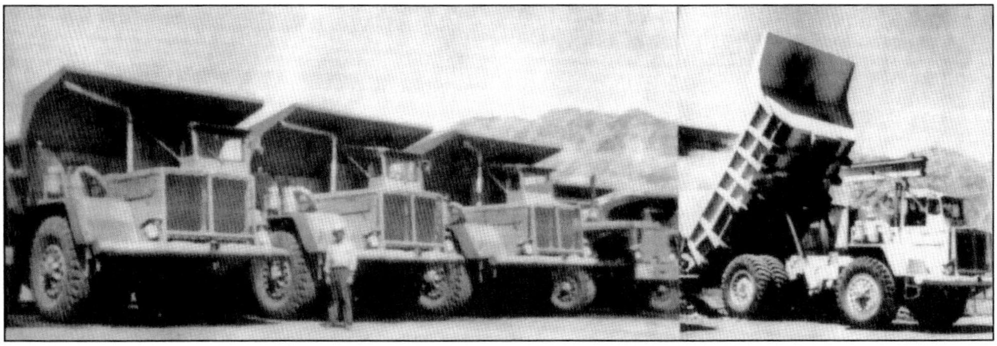

Much of the ore at Eagle Mountain Mine was hauled from the pit to the crusher in three Kenworth 40-ton, spring-mounted end-dump trucks. These units were thought to be the largest two-axle trucks produced at the time. Designed specifically for off-highway hauling jobs in open-pit mining and earth-moving industries, the empty trucks weighed 34 tons. They were 13 feet high, 14 feet wide, and 24 feet long. The mounted tires were six feet in diameter. The trucks were powered by a single 300-horsepower, turbocharged diesel engine. It had a short turning radius, relatively light chassis weight, economy of operation, and ease of maintenance. The bodies were lined with approximately 5,000 pounds of Kaiserloy steel plates that were three-quarters-of-an-inch thick. The truck featured power steering and "Torqmatic" transmission and brakes. The truck hoist was a patented Kenworth three-stage, two-cylinder telescoping mechanism. The trucks at Eagle Mountain Mine were the first of their type used in the mining industry.

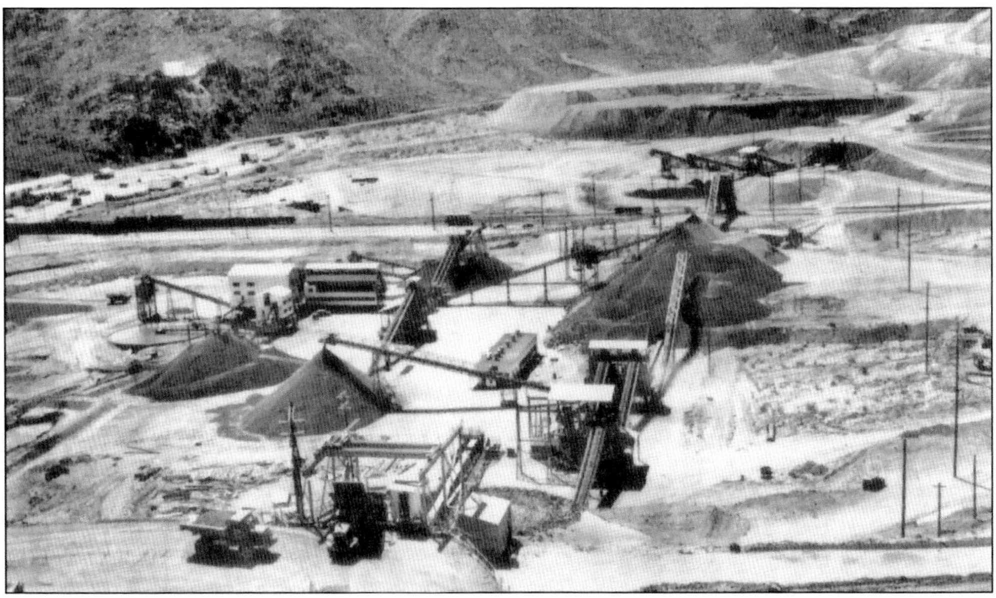

Big power shovels attacked the broken rock and loaded it onto trucks that hauled it to the mine's crushing plant. These huge diesel-powered shovels at the Vulcan Iron Ore Mine in Eastern San Bernardino County worked on the 50-foot benches of an open pit. Blasting broke the ore from the earth, and the shovels lifted two-and-a-half yards of ore with each bite, dropping it into diesel trucks. The ore was then hauled to a crusher at the mine edge, where it is broken down into 10-inch chunks and delivered by conveyor belt to storage piles.

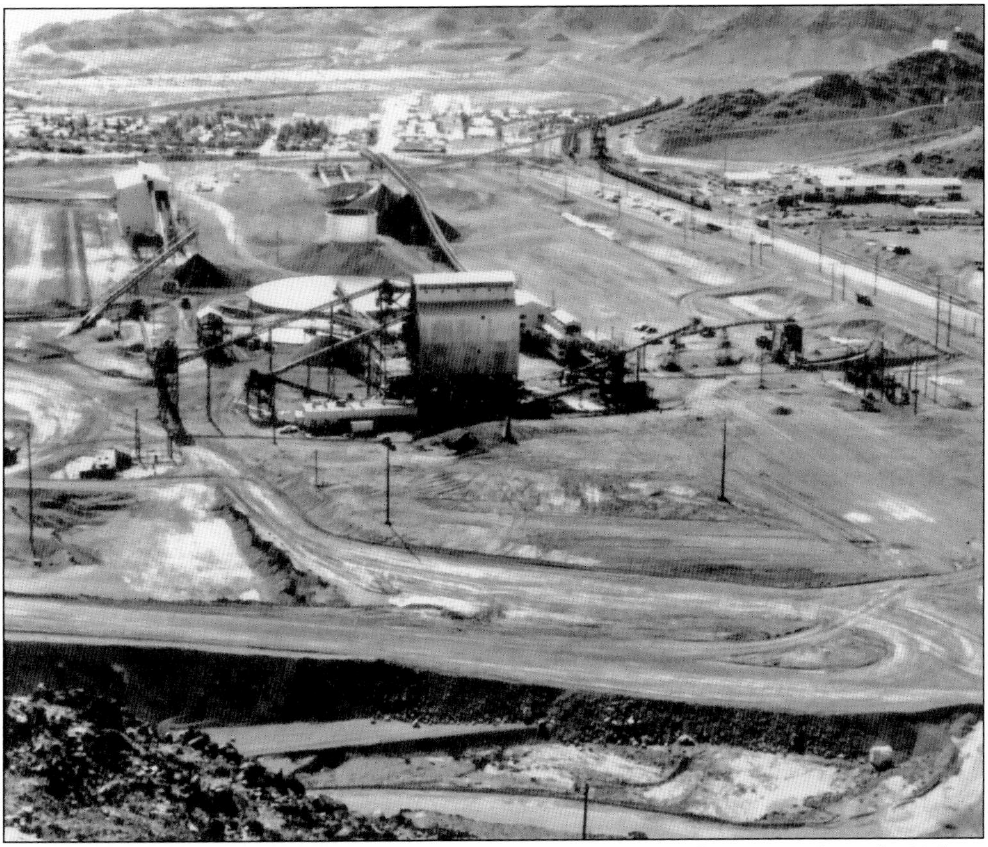

Mammoth diesel trucks shuttled the ore to the railroad. Although all ore came from the Vulcan Mine, there were, and are, numerous other high-grade deposits within a 300-mile radius of the Kaiser Mill, which kept it in continuous operation for decades. Shipments of 1.599 million tons of iron ore were made during an average year from the mine during the mill's heyday.

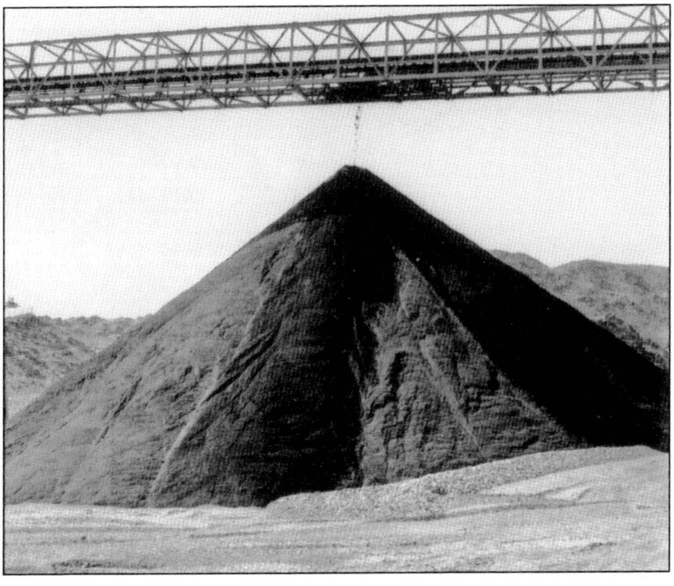

The ore is then conveyed to the Eagle Mountain beneficiation plant. Beneficiation is the method of separating the richest ore deposits from waste materials and ore with low iron content before shipments to the steel mill. The ore prior to installation of the beneficiation plant contained approximately 51 percent iron. Beneficiation raised the content to almost 60 percent, increasing blast-furnace capacity at Fontana.

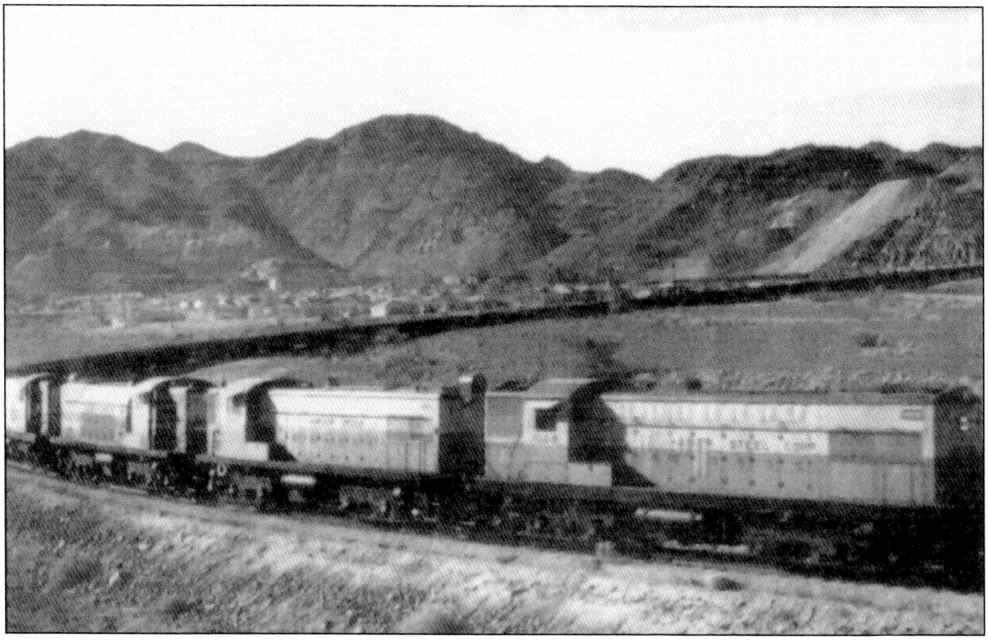

The mine was 164 miles from the Fontana Steel Plant, making possible one of the shortest iron-ore hauls in the steel industry. The first 52 miles are over a standard-gauge railroad, built and operated by the company. It traversed desert to meet the Southern Pacific main line at Ferrum, on the shores of the Salton Sea. From this junction, the ore cars—140 to 160 fully loaded gondolas daily—were hauled the remaining 112 miles to the mill. Southern Pacific engineer John Rapoza ran trains hauling ore on this route using three engines in front, three in the center, and three at the rear for at least 100 fully loaded cars. He is retired. His counterpart was Harold L. Hostetler, who worked from July 1944 to November 1956, before transferring to the tin-mill lab. Hostetler passed away in December 1983.

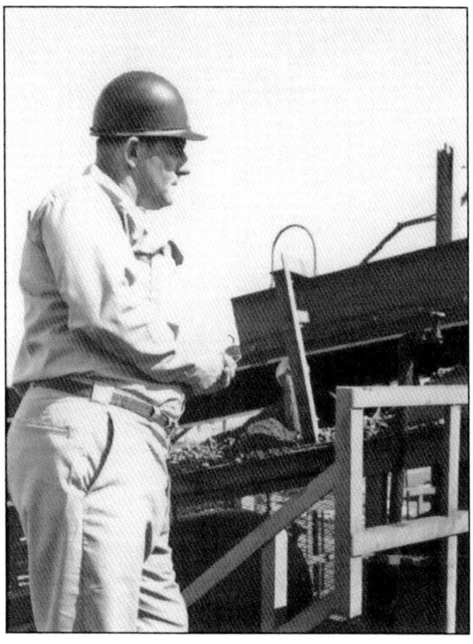

Victor "Vic" Fischer (pictured) was hired in 1950 as a railroad engineer at Eagle Mountain. In 1952, he became assistant superintendent of the railroad. Around 1953, he was promoted to superintendent when Ben Myer retired. He made many improvements on the miles of track between Eagle Mountain and the Salton Sea. He also was responsible for newly designed ore cars for transporting ore from Eagle Mountain to Fontana Steel Mill. Kaiser Steel owned an interest in Hammersley Steel in Dampier, West Australia. Shipping output was low and the operation was not being run as efficiently as Kaiser would have liked. In 1967, Vic was sent to Australia to make a general overhaul of the operation at Hammersley. After a few months, efficiency and output were greatly improved. Vic spent about a year in Dampier before returning to Eagle Mountain where he continued as railroad superintendent. He retired in the early 1970s.

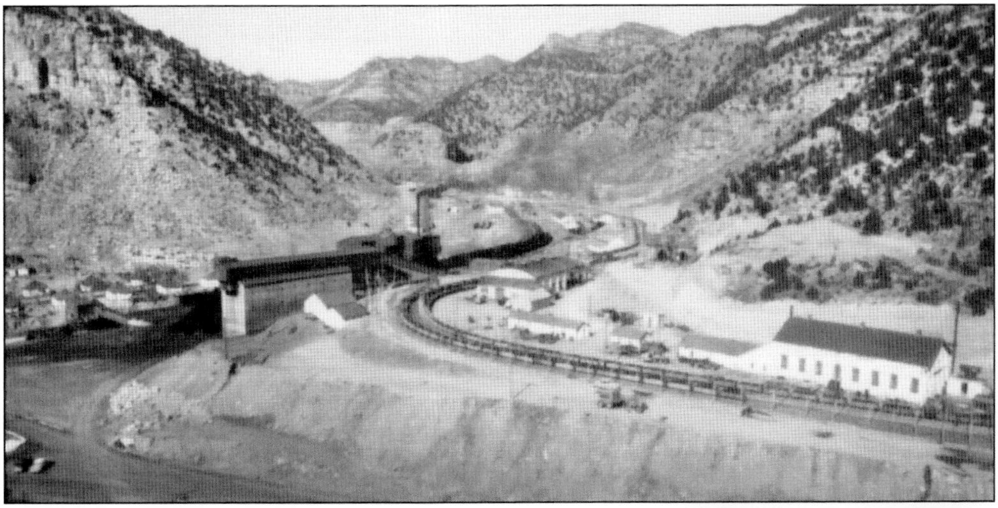

The Sunnyside Mines of the Utah Fuel Company were located in southeastern Utah and were leased and later bought by Kaiser Company, Inc. These deposits assured a long-range supply of high-quality coking coal, rich in by-products. Here miners are working two miles under the picturesque Book Cliffs, blasting out some of the 615,000 tons of coal needed annually for the Fontana coke ovens. Crushing, blending, washing, and dewatering (to remove impurities) the coal were accomplished at the mine in a modern jigging and floatation plant. Pulverization was the only step remaining before the coal was ready for charging in coke ovens at the Fontana Steel Mill. During some years, more than one million tons of coal were shipped to Fontana. The 297 beehive coke ovens operated at full capacity seven days a week. More than 750 miners lived at Sunnyside with their families in a modern community. The company also operated a large coal mine at York Canyon, New Mexico.

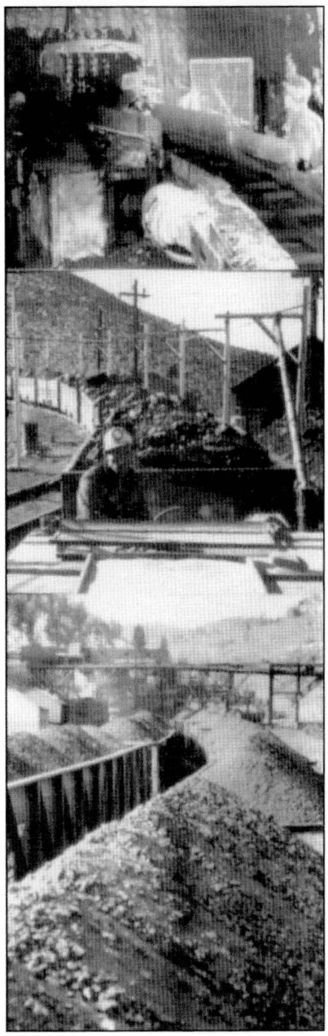

In July 1955, Kaiser bought all of the real estate and mining properties of the St. Louis, Rocky Mountain and Pacific Company at Raton, New Mexico. The company acquired title to 203,000 acres of coal-bearing land and the coal-mining rights to an additional 327,000 acres. Shipments of coked coal have been made to Fontana, where the high quality of Raton coal has been proven by actual use.

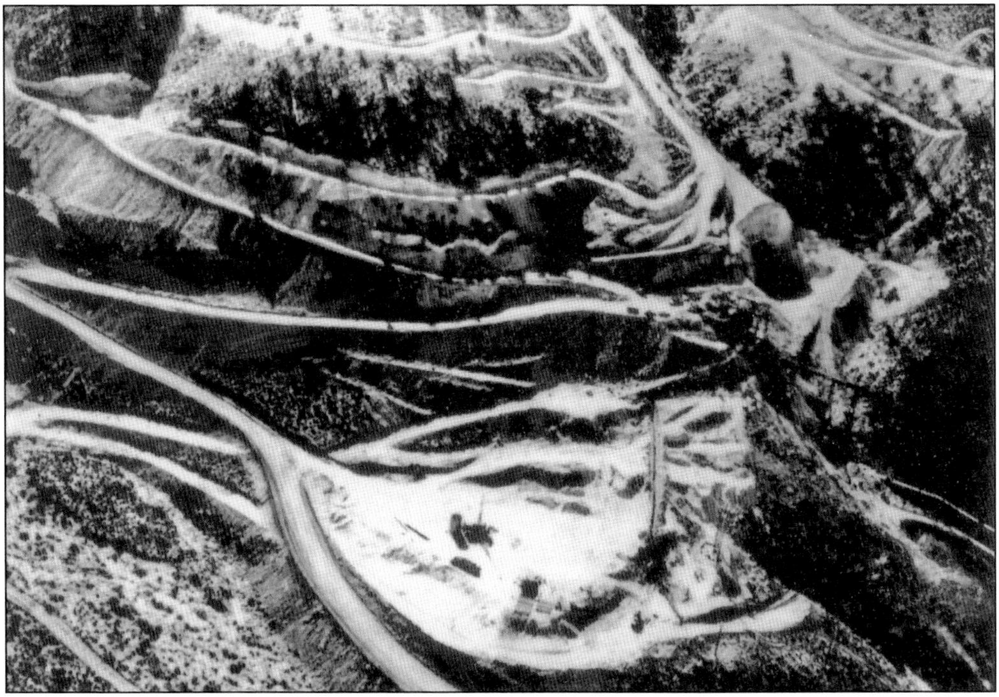

Used as flux in the open-hearth furnace, dolomite came from Sloan, Nevada, and Salinas, California. Sufficient deposits of manganese ore were already developed in the Southland to supply the needs of this metal. Another essential raw material needed for the production of pig-iron in Kaiser Steel's furnace was limestone. The mining of the limestone was handled by Permanente Cement Company, an affiliated Kaiser industry, which operated a cement plant at Cushenbury. On September 30, 1955, the company bought a large high-grade deposit of limestone at Cushenbury, California, only 75 miles from the Fontana plant. Limestone had been purchased from Nevada. Actual testing of Cushenbury limestone in Fontana blast furnaces and open-hearth furnaces proved the high quality of the limestone.

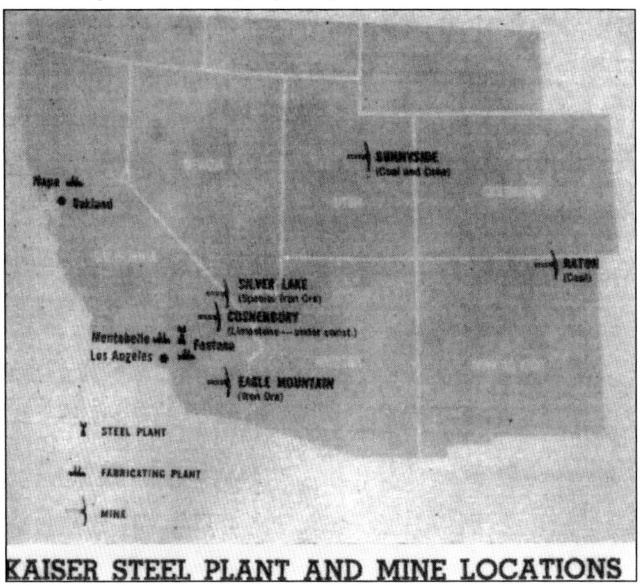

This is a map of locations for supplies of raw materials. The limestone property was located near the town of Lucerne Valley, 30 miles southeast of Victorville, California. The Atchison, Topeka & Santa Fe Railway agreed to build a 32-mile railroad connecting the quarry with the main line at a point near Hesperia, California. The acquisition of this limestone deposit made Kaiser Steel completely self-sufficient of the raw materials needed for the operation of its three blast furnaces.

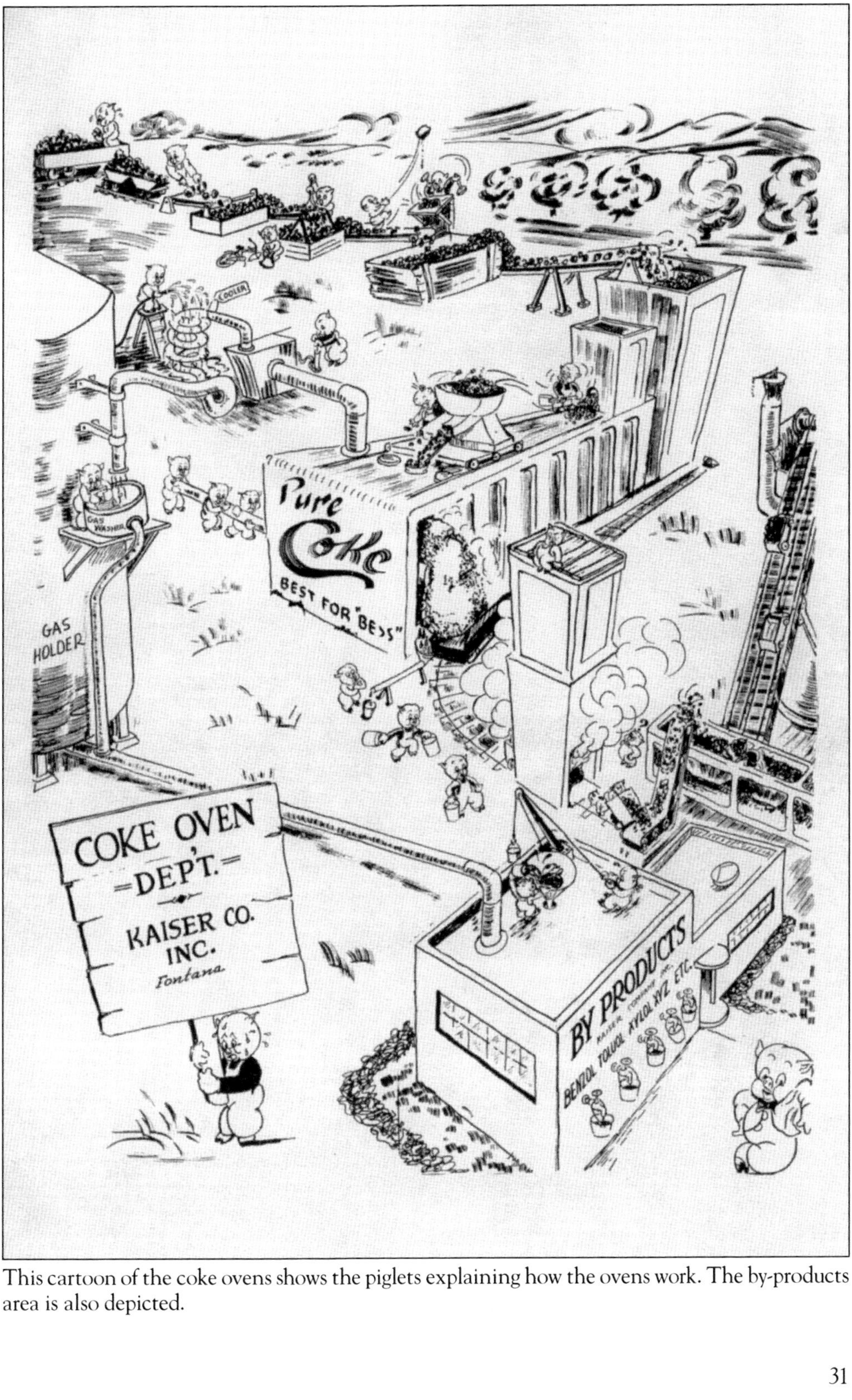

This cartoon of the coke ovens shows the piglets explaining how the ovens work. The by-products area is also depicted.

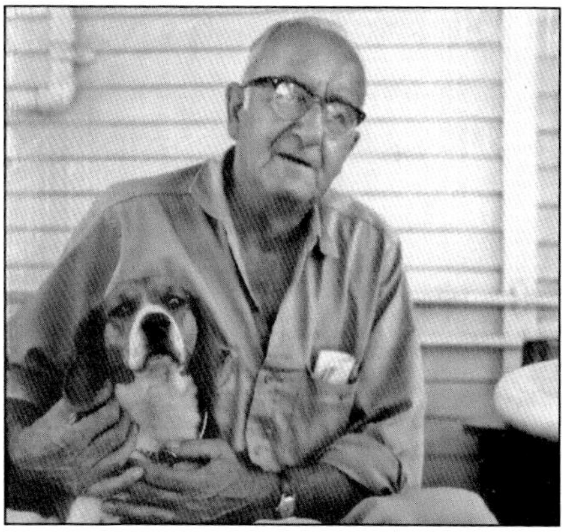

Russell "Slim" Putnam started work at Kaiser in December 1942 and retired in 1958, putting in 16 years of work. He established the conveyor-belt shop at Kaiser Steel, where the belts where made. A traveling bin, known as a lorry car, filled each oven by drawing 14 tons of coal from the storage tower and dropping it through three charging holes in the top of each oven. The charge is then evenly spread with a leveling bar on a "pusher," a huge machine running on rails on the north side of the ovens.

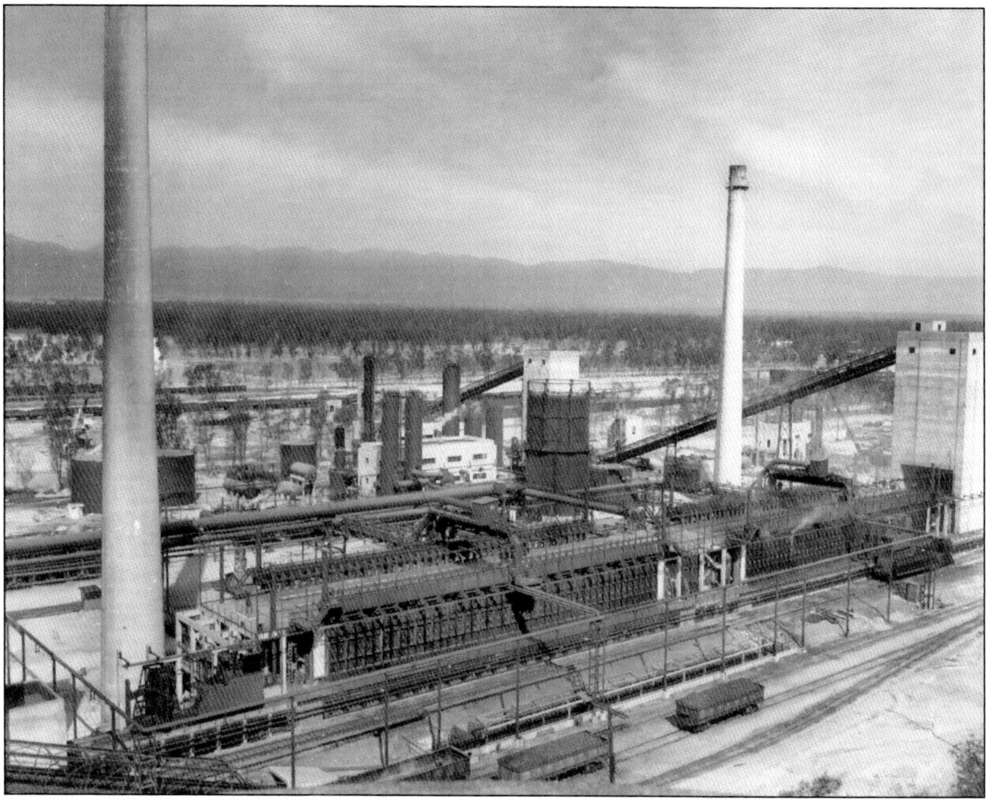

This aerial image of the coke ovens, looking northeast, was taken on February 27, 1943. The carboning process then started, and the 14 tons of coal were scientifically heated for 18 hours, driving off gas or volatile matter that is sent by pipeline to the by-products plant nearby. The coke-oven gas passes through the recovery apparatus where the various by-products are recovered. A portion of this gas is used for oven heating and the surplus fuel throughout the plant. With clockwork regularity, 24 hours each day, a slab of coke 14 inches wide, 40 feet long and 14 feet high was pushed at 12-minute intervals from one of the 90 coke ovens at the steel mill. This may appear to be a simple process, but the operation of a by-product coke plant is an intricate, highly scientific process. The Kaiser Company's Koppers-Becker under-jet coke ovens consisted of two batteries of 45 ovens each, having a combined capacity to produce 935 tons of coke from 1,728 tons of coal every 24 hours. (Courtesy of Alexia King.)

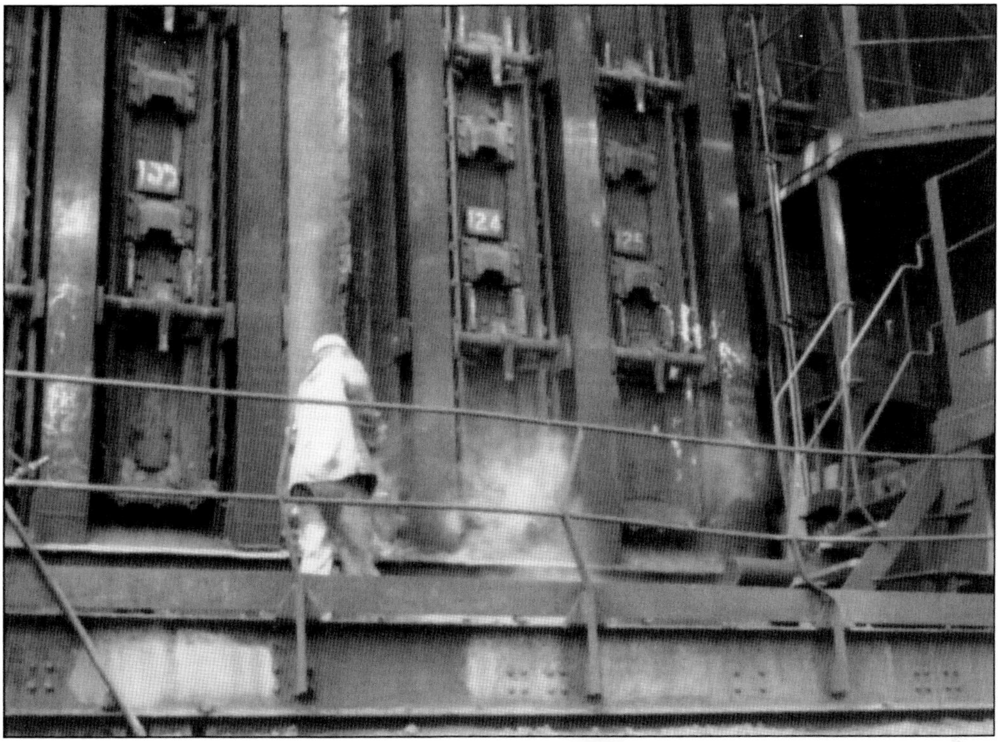

This is a 1977 coke oven with the oven worker overseeing the "pusher." At the end of the previously described 18-hour period, the ram on the "pusher" forced out the charge of coke, which glows like a cherry-red honeycomb. The coke is then cooled in a quenching tower, where streams of water spray over the glowing mass and clouds of steam billowed skyward. The coke was then screened and graded for use in the blast furnace. (Courtesy of Charles Dall.)

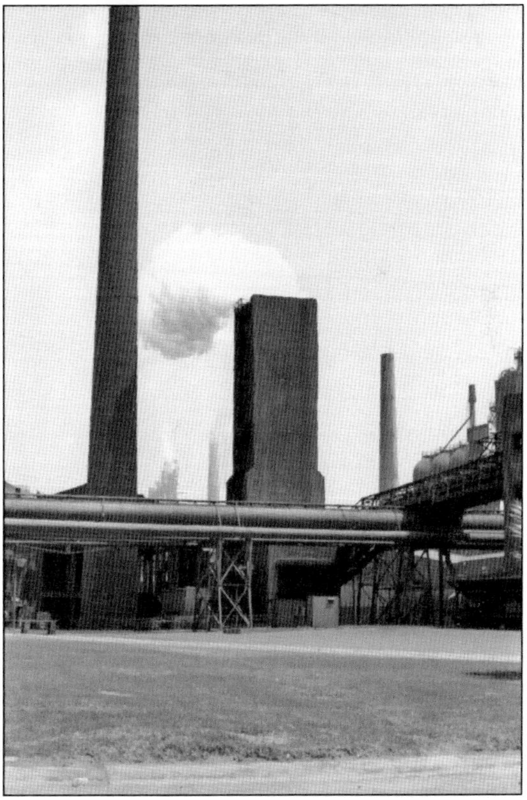

Steam coming out of the quenching tower is shown here in August 1970. (Courtesy of Charles Dall.)

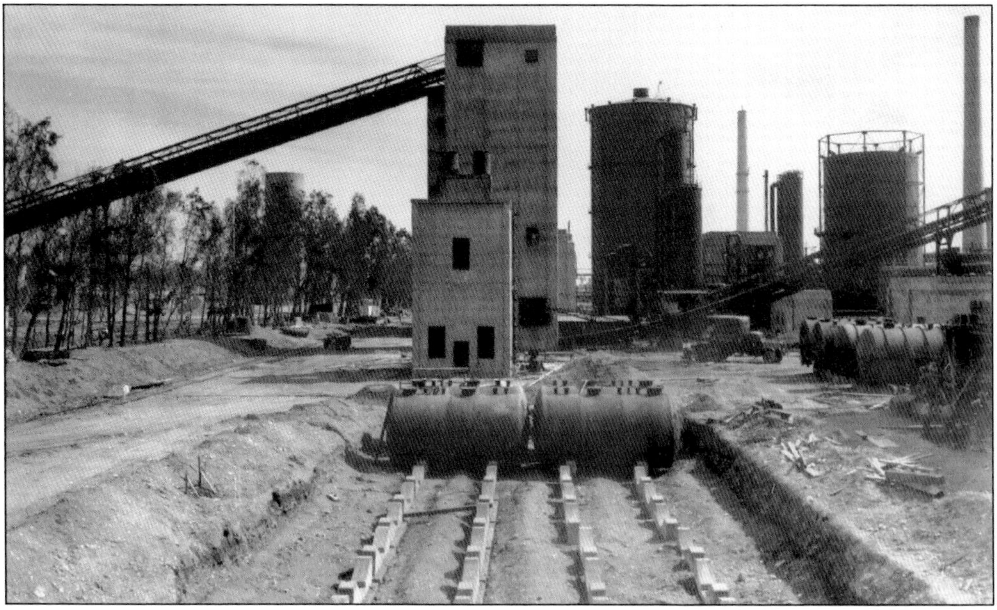

This is the coke plant and the by-products area under construction in February 1943. The Benzol plant and underground storage tanks are at center in the foreground. To continue the coking process, the volatile matter, which has been driven off, was refined in the by-products plant, where coal tar, ammonium sulphate, sodium phenolate, light oils, creosote oil, pitch, and residue oil were reclaimed. These basic by-products were refined further, into literally hundreds of products used by men and women in everyday life, ranging from delicate perfumes to powerful explosives. (Courtesy of Alexia King.)

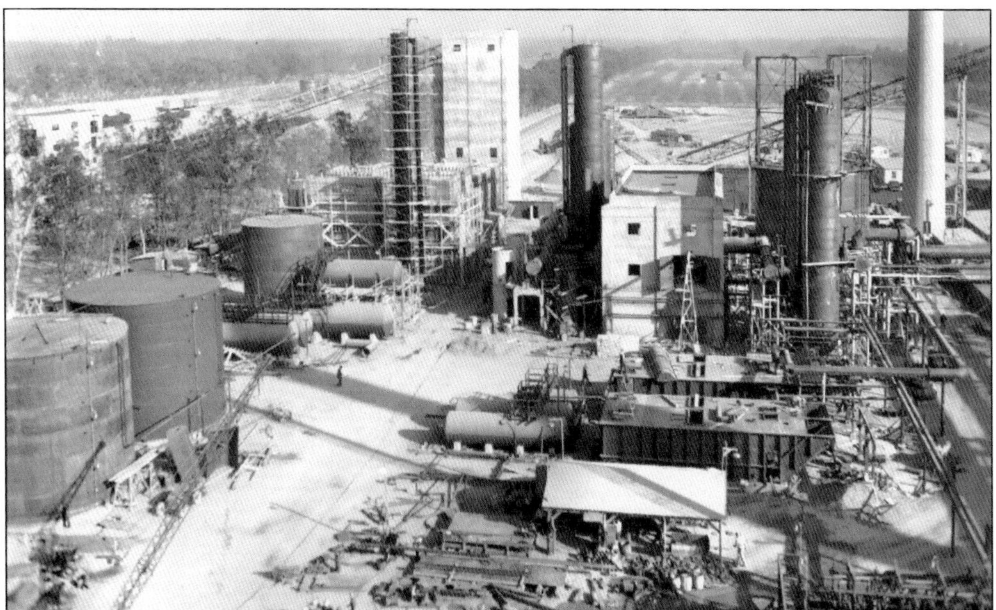

This is a by-products area after construction in 1943. The coke ovens and the coking process are in themselves practically an individual industry and are, in fact, a vital part of steel manufacturing. (Courtesy of Alexia King.)

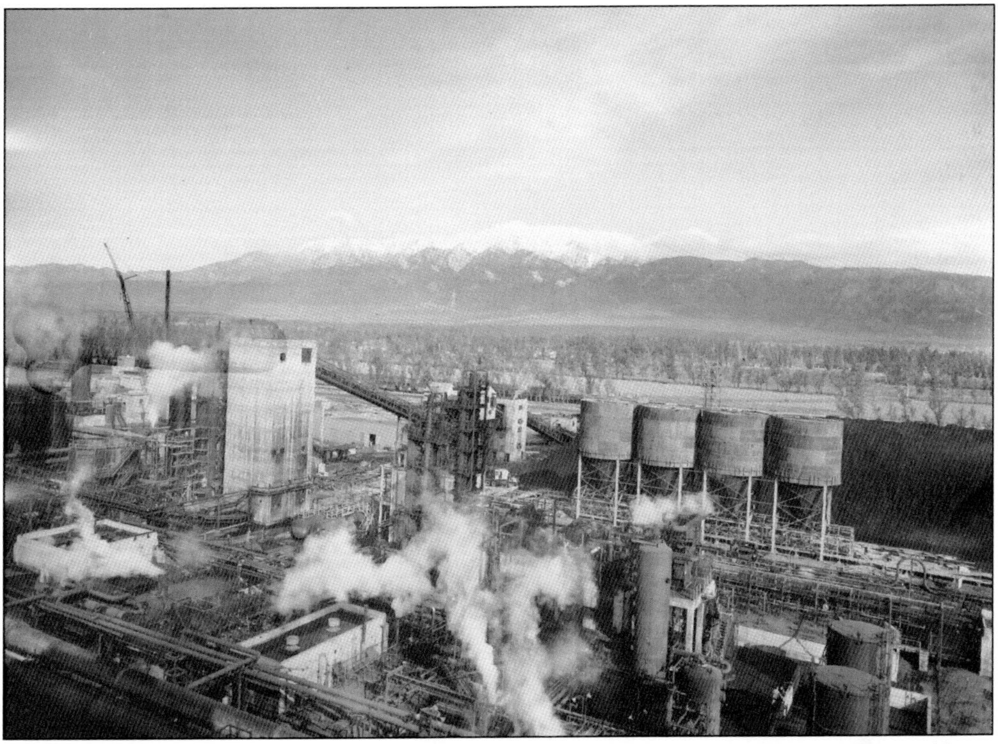

The by-products area, the coke ovens, and the nearby handling breaker building are pictured here. This image was taken in February 20, 1958, with new snowfall on the San Gabriel Mountains north of the plant. (Courtesy of Jonnie Long.)

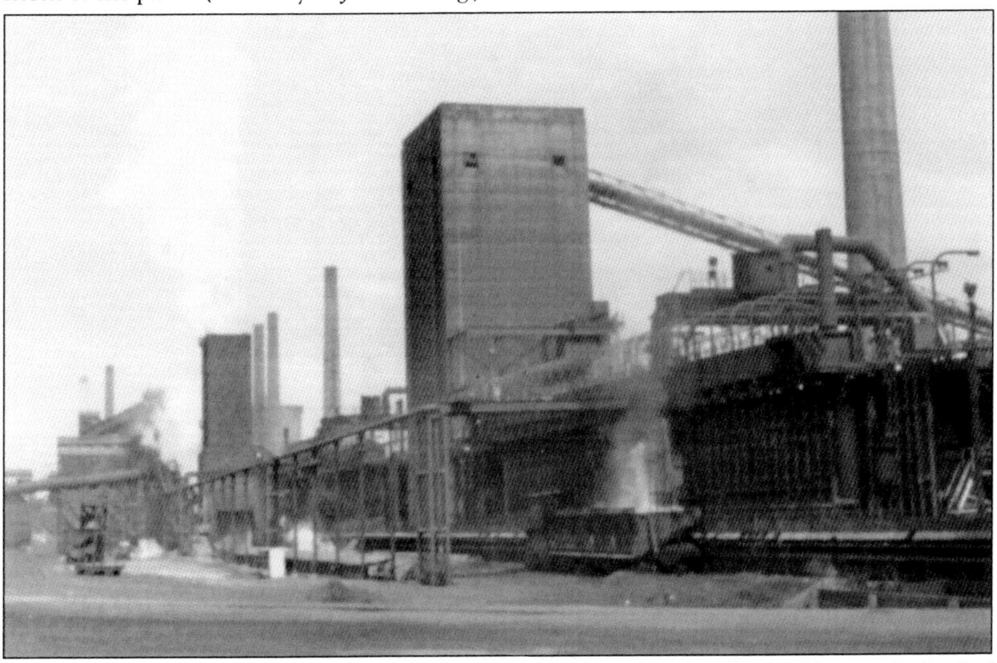

This is another photograph of the coke ovens with the handling breaker building and a load of coke coming out of the oven into a railroad car. (Courtesy of Charles Dall.)

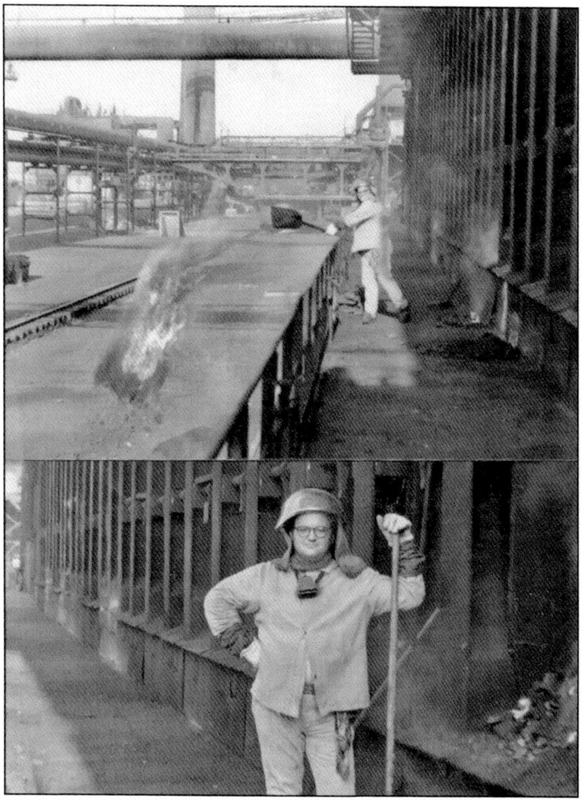

Steven Scalf is pictured working at the coke oven with the tools of the job and the protective gear and clothing he had to wear, no matter what the weather. He worked for 12 years, having arrived in Fontana from the state of Indiana. (Courtesy of Steven Scalf.)

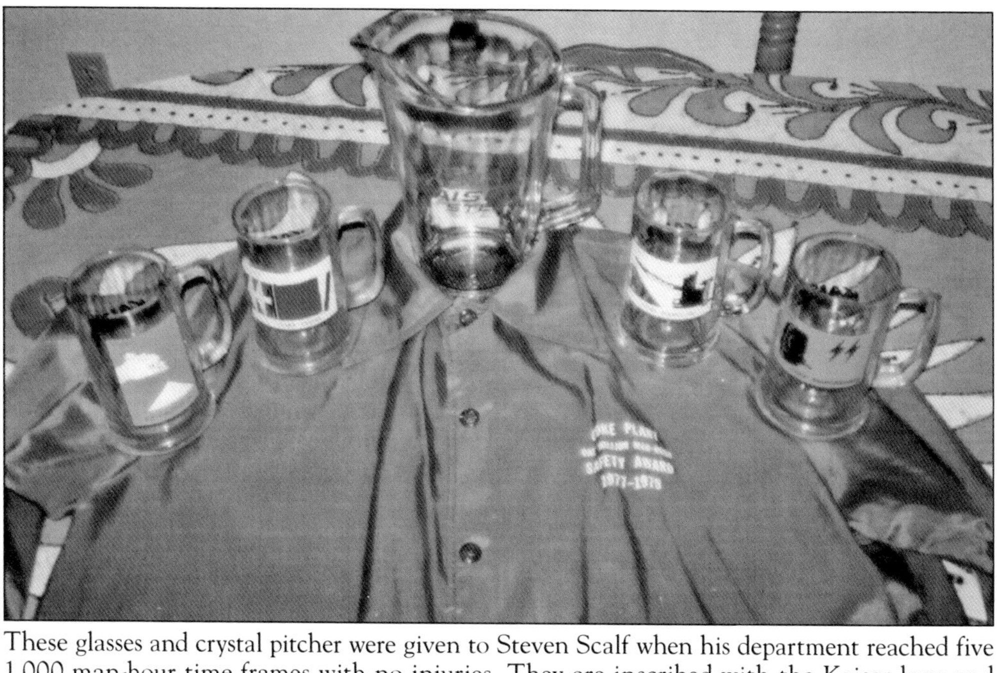

These glasses and crystal pitcher were given to Steven Scalf when his department reached five 1,000 man-hour time frames with no injuries. They are inscribed with the Kaiser logo and department name and are prized artifacts. (Courtesy of John Anicic Jr.)

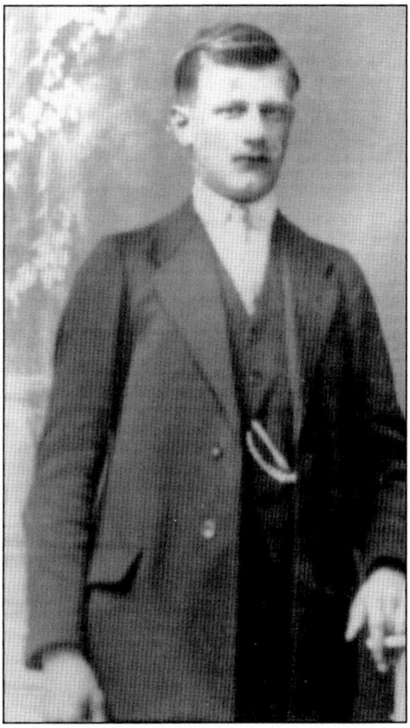

Frank Videgar (left) worked in the by-products area for 15 years and came from the state of New York. His son William "Bill" Videgar (right) worked in the coke oven for 34 years after arriving in Fontana from the state of New York. He worked the last coke-oven shift, as the lead pusher, of the operation on December 29, 1982. The button was pushed at 11:00 a.m. sharp on Oven No. 48. (Courtesy of William Videgar.)

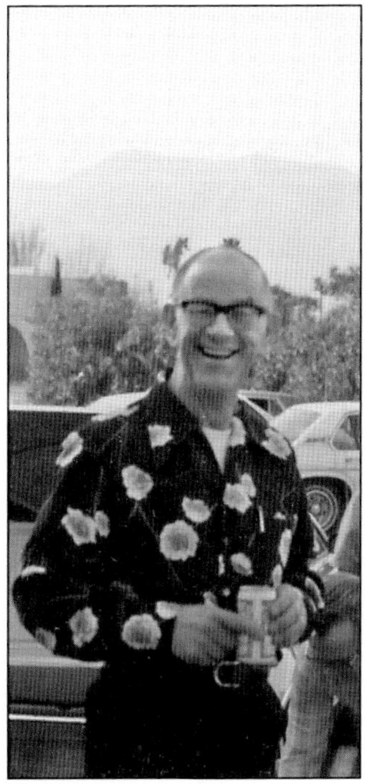

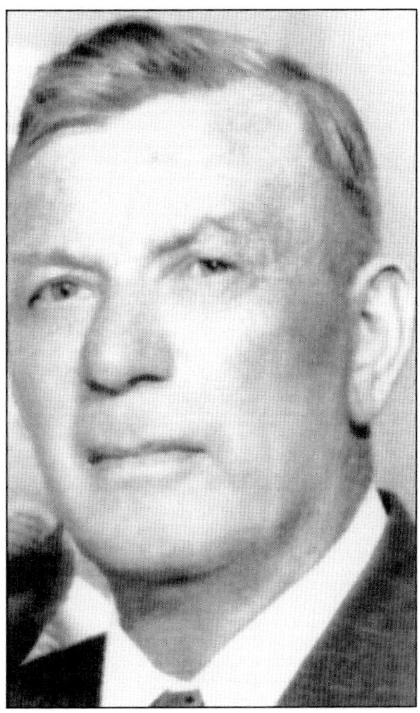

Louis Lipovec worked in the by-products area. He was a pitch-machine operator at Kaiser for 10 years. Lipovec previously worked in mines throughout the West and in Minnesota.

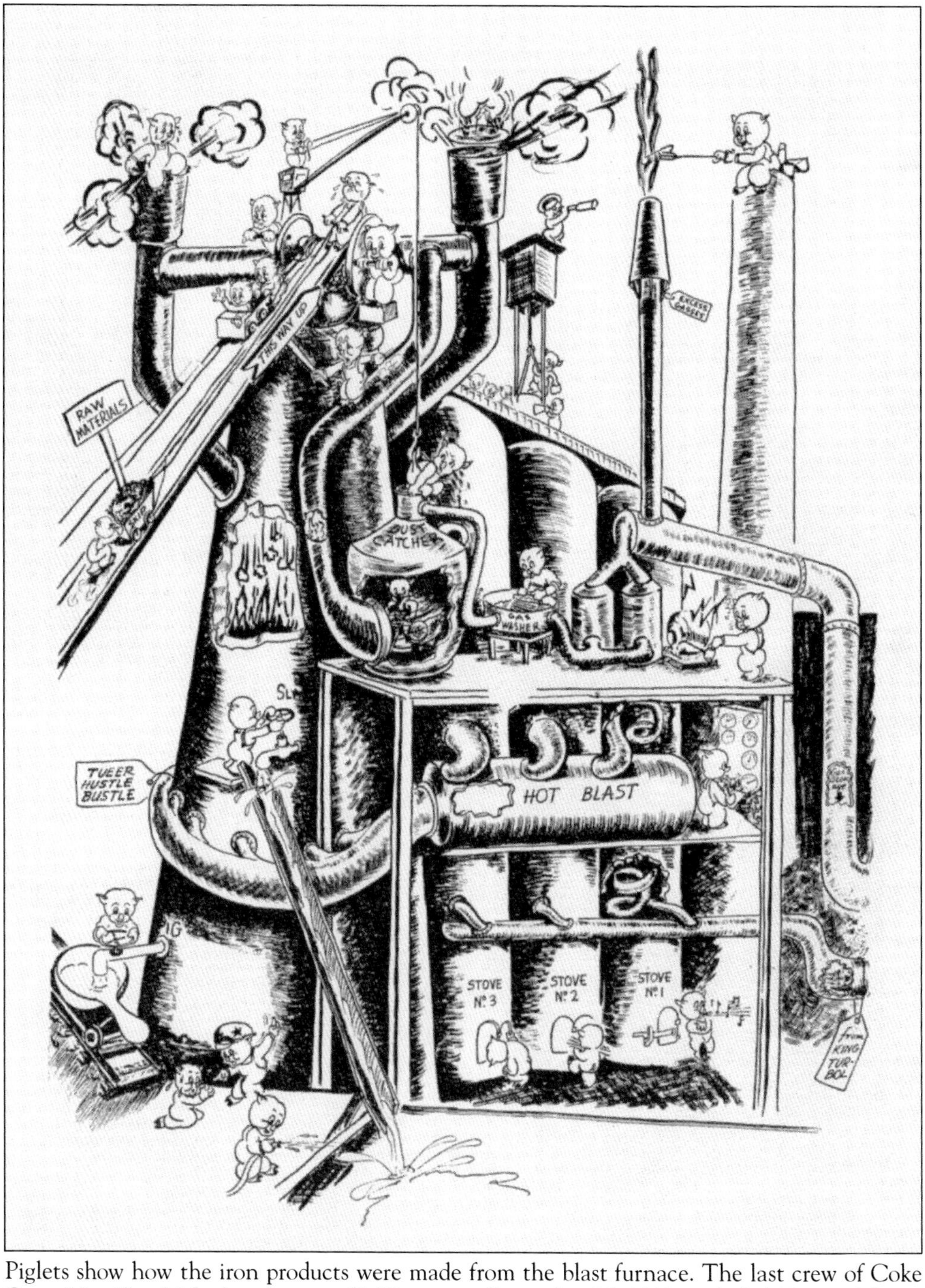

Piglets show how the iron products were made from the blast furnace. The last crew of Coke Oven No. 48, which shut down at 11:00 a.m. on December 29, 1982, included Henry Adams, Paul (Marble Eyes) Alanis, Turkey-Neck Bill, Sugar Carter, R. Cunningham, Richard (Dick) Dominquez, Roy Gagne, Pee Wee Marshall, Jesus Navarro Sr., Tony Rodriquez, Richard Sosa, Ralph Torango, Angie Vantola, William (Bill) Videgar, Jim Wharey, Pete Zeegers, and Tom Zuiroz. The foreman was John De Sort, and the superintendent was Clarence Benninghof. Les Price also worked in the by-products area.

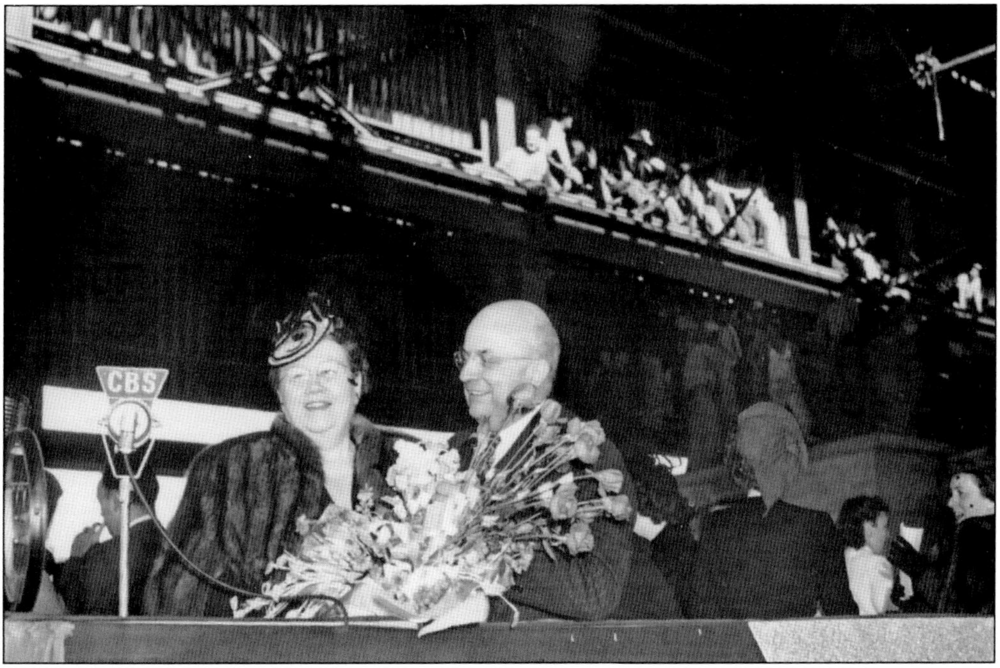

Mr. and Mrs. Henry J. Kaiser are pictured on December 30, 1942, at the dedication of the first blast furnace, built in eight months. Mrs. Kaiser's name is Bess, and her husband named the first blast furnace after her. The plant was now ready to start producing steel for the war effort. A large crowd was in attendance at the dedication. Twelve hundred tons of pig iron every 24 hours was the estimated production capacity of "The Bess."

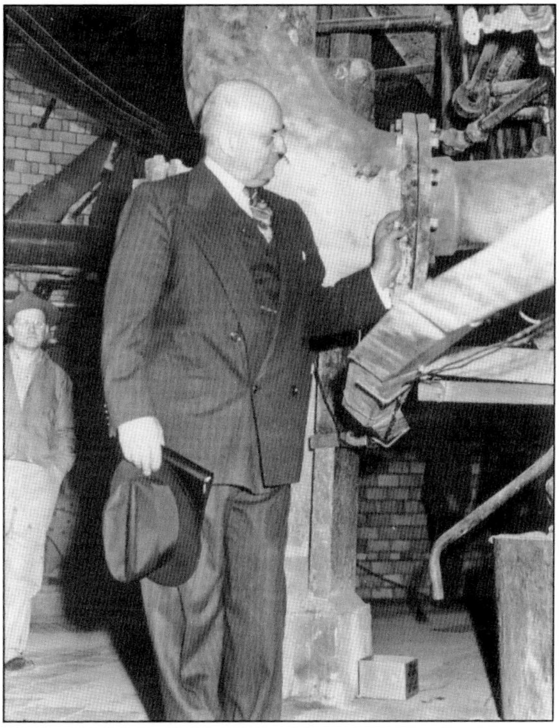

Henry J. Kaiser looks over the blast furnace after the dedication. Twelve hundred tons of pig iron can be manufactured into an infinite variety of products. For instance, should one day's production be used for the manufacture of electric irons, there would be enough metal to make an iron for every man, woman, and child in Arizona. Or, if made into ships, there would almost be enough iron to make enough steel to build a complete Liberty Ship every other day. "The Bess" looked like a huge steel milk bottle, measuring 130 feet around at the base and towering over 100 feet into the sky. It was lined with three feet of expensive, high-heat-resisting brick, into which the ingredients for pig iron were poured continuously.

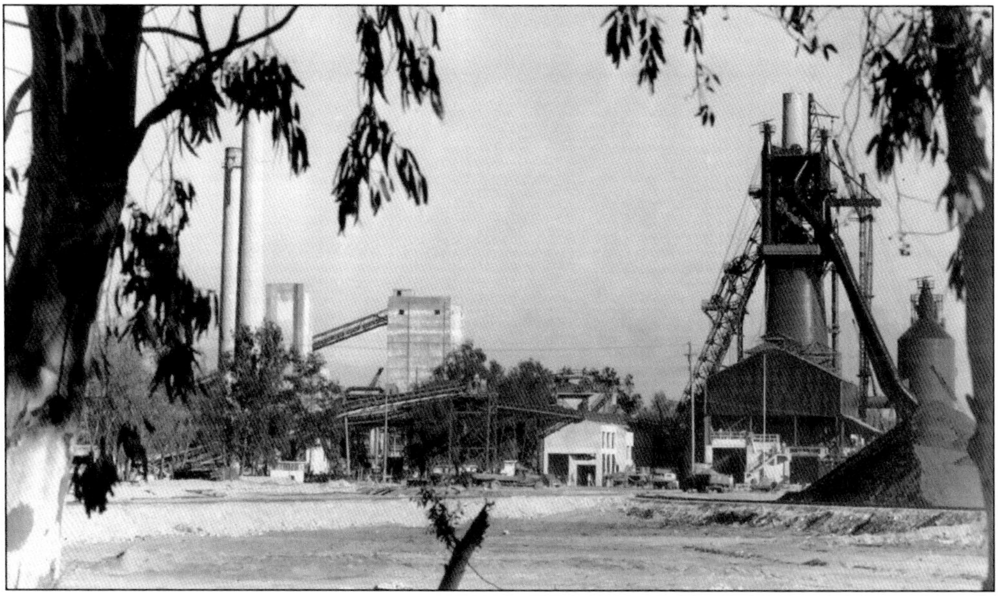

A view through the trees around the plant shows the blast furnace. The "charge" of coke, limestone, and iron ore is melted by a hurricane-like blast of air, preheated in three hot blast stoves at the side of the furnace. Burning blast-furnace gas, these stoves generated intense heat in excess of 1,800 degrees, the required temperature for the blasting process. At the powerhouse nearby, a powerful turbo-blower forced air at the rate of 100,000 cubic feet per minute into the stoves, where it absorbed the heat and was then forced directly into the bosh of the blast furnace. The combination of this hot blast and the burning coke melts the charge within the furnace into pig iron and slag. (Courtesy of Alexia King.)

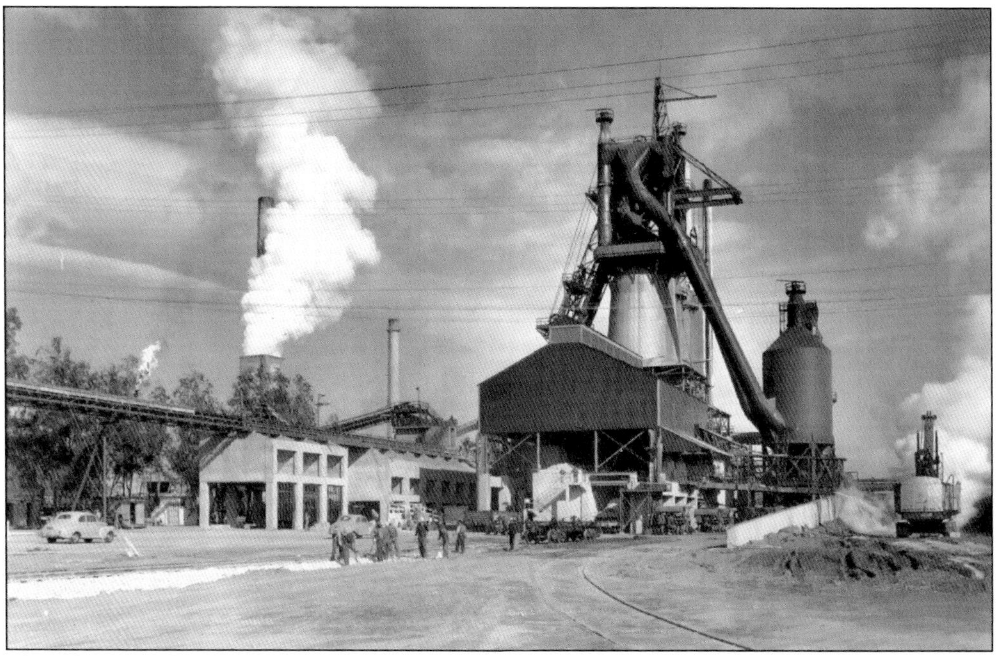

This close-up of "The Bess," looking northeast, was taken on February 3, 1943. Alphos "Al" Mendoza was one of the firemen who worked the blast furnace. (Courtesy of Alexia King.)

The No. 2 blast furnace is pictured here blowing, or "snorting" as it was known, in 1969. The hot blast suspends the "charge" within the furnace and drives off gases at the top with a velocity of six miles per minute. These gases were caught, cleaned, and used to reheat the hot blast stoves. In making a single ton of pig iron, four tons of air, by weight, must be used. The air is blown in through tuyeres, or holes in the side of the furnace, at a maximum rate of one million cubic feet per minute. (Courtesy of Charles Dall.)

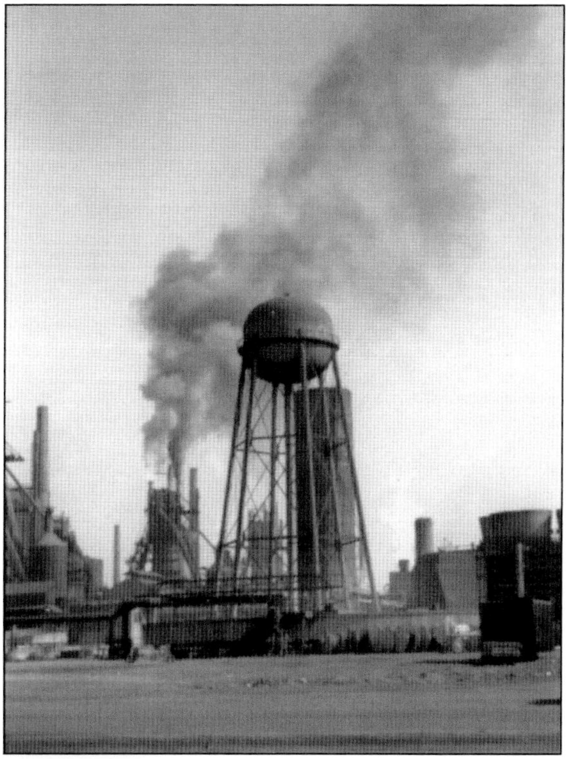

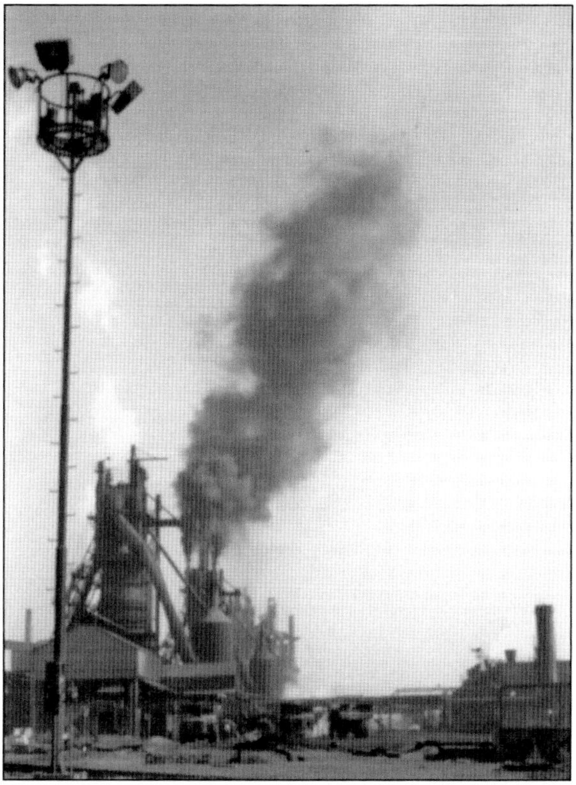

This is another view of the blast furnace blowing, or "snorting," taken in 1969. (Courtesy of Charles Dall.)

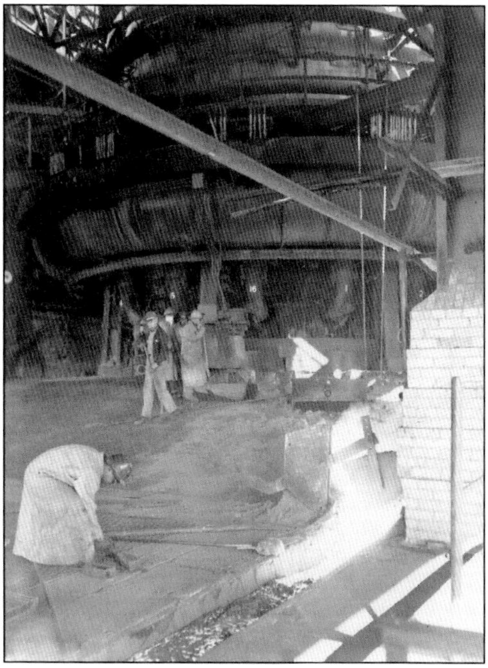

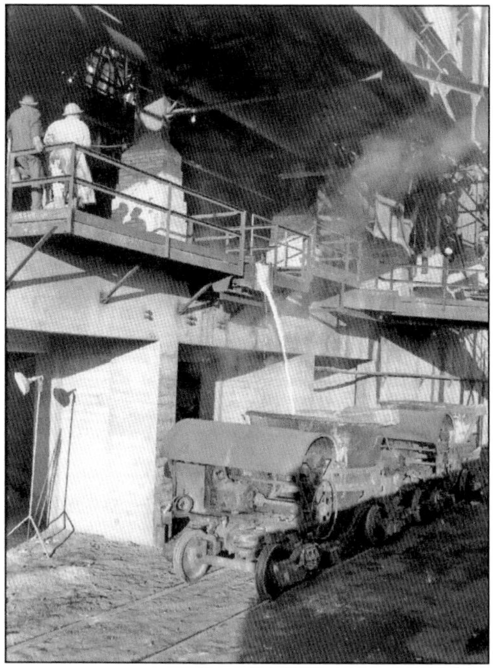

Tapping of the iron from the blast furnace is depicted in this February 3, 1943, photograph (left). As molten pig iron melts to the bottom, the furnace was tapped and the iron and slag was drawn off. The slag was poured into a concrete pit, later to be sold for use in concrete aggregate, insulating material, road building, rock wool products, and many other varied uses. The pig iron was drawn into 150-ton hot ladle cars and transported either to a casting machine where it was poured into "pigs" to be sold to manufacturers or taken directly to the open hearth in a molten state to be made into steel. A photograph from the same day (right) shows tapping of the blast furnace and molten iron running into the hot-ladle car. (Courtesy of Alexia King.)

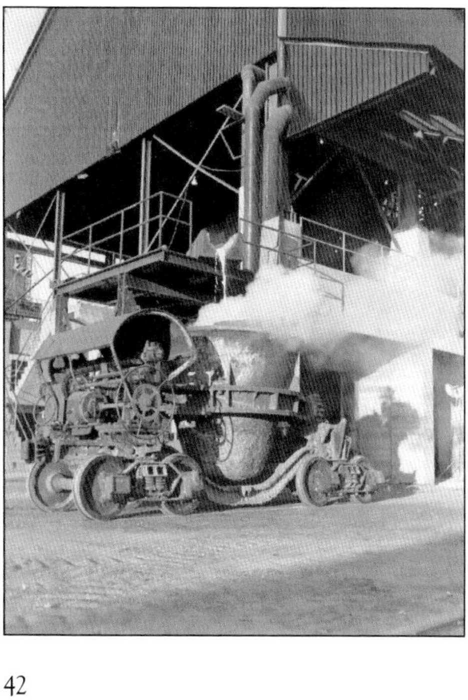

This close-up of the ladle car with iron running into it was taken on February 3, 1943. (Courtesy of Alexia King.)

The blast furnace is pictured here. (Courtesy of Alexia King.)

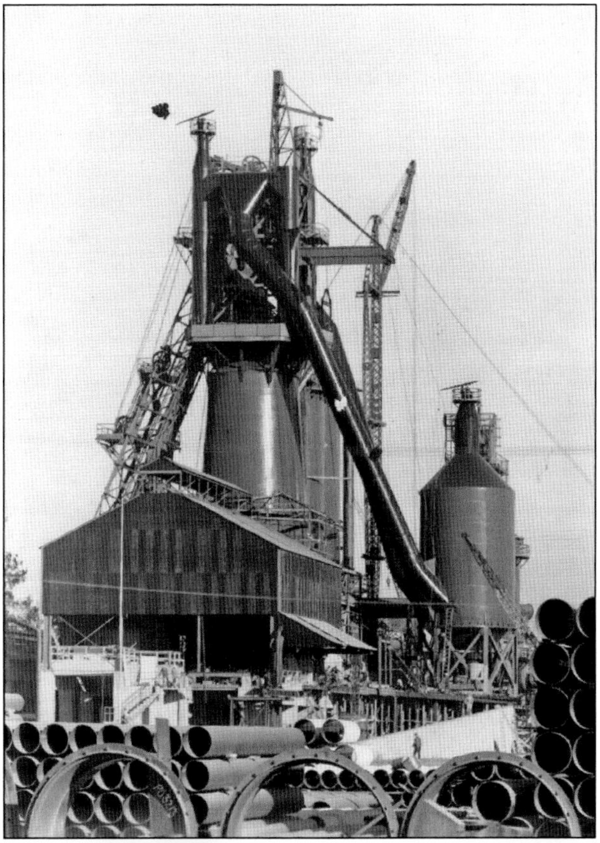

A view of the blast furnace, looking southeast, also shows the railroad cars with materials in them. It was taken on February 3, 1943. (Courtesy of Alexia King.)

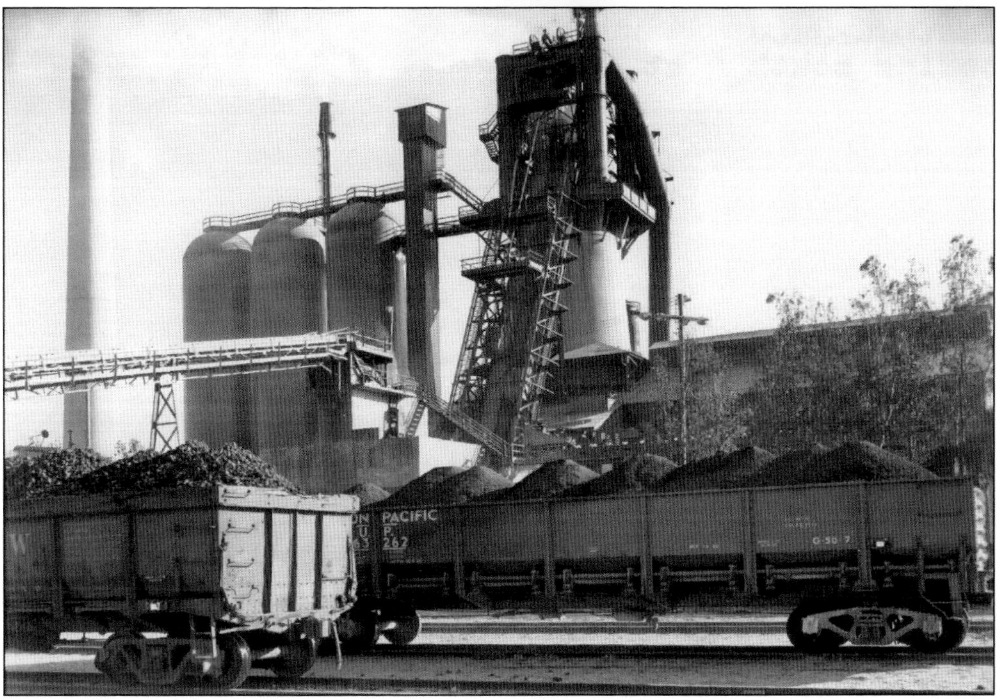

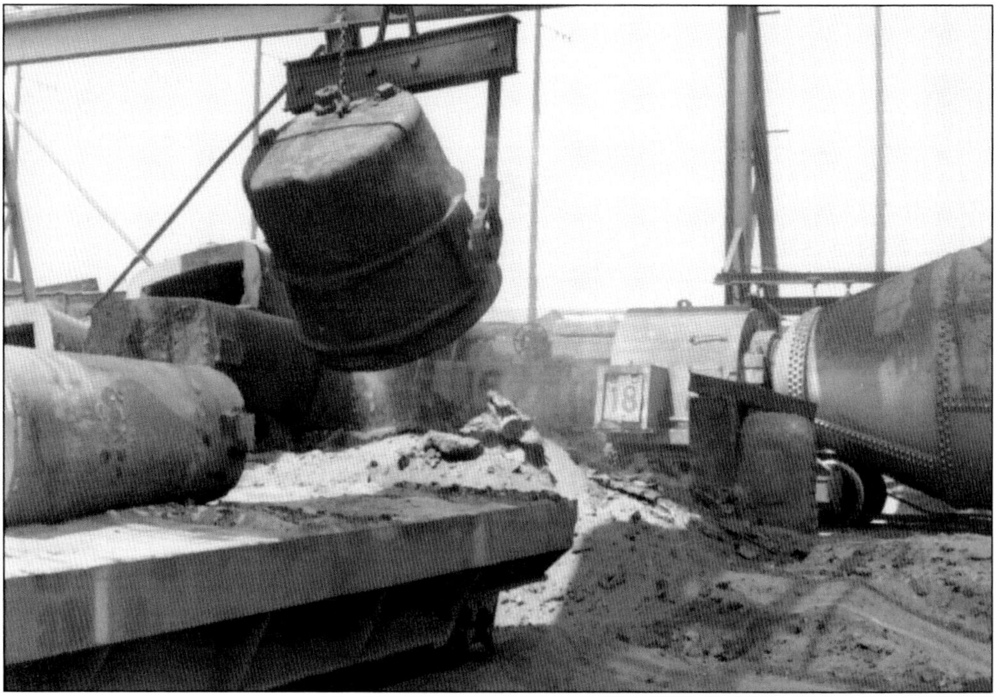

This is a view of the northwest end of the mill foundry. (Courtesy of Charles Dall.)

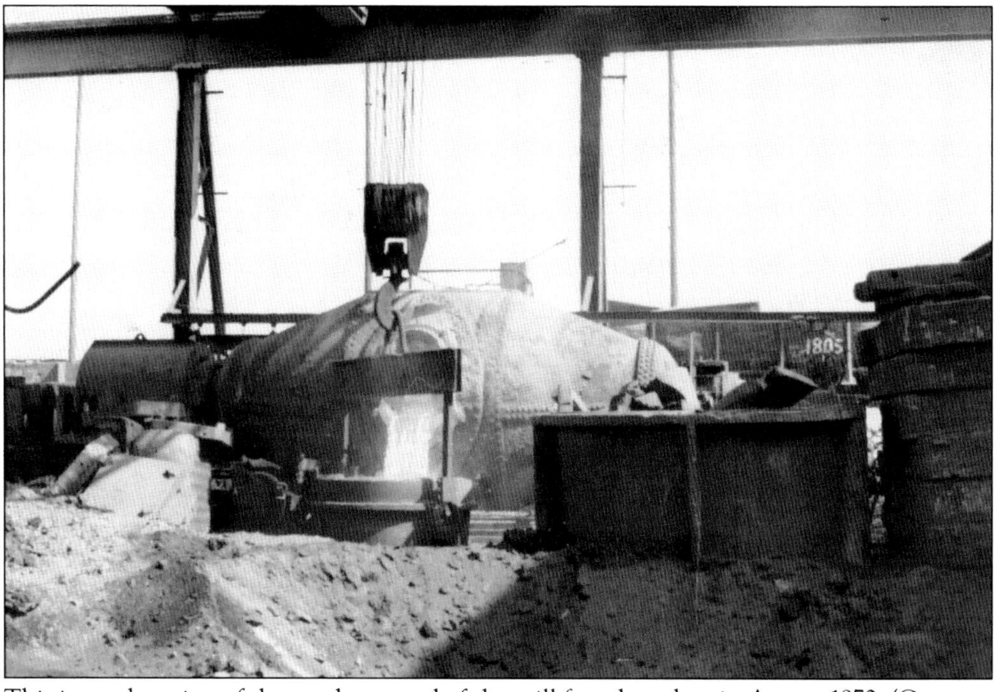

This is another view of the northwest end of the mill foundry taken in August 1970. (Courtesy of Charles Dall.)

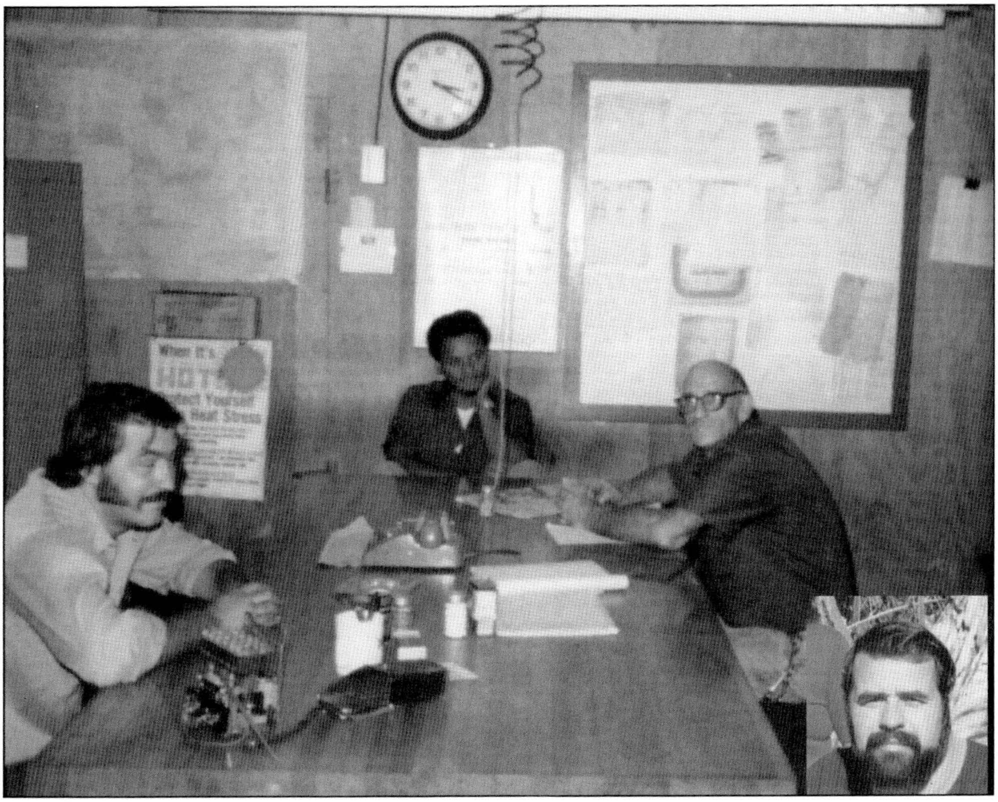

Pictured here in the new oxygen shop break room, from left to right, are Ray ?, Henry Brown, Carlos Zapata, and Richard (Dick) Bennett (inset), who worked for 19 years at Kaiser.

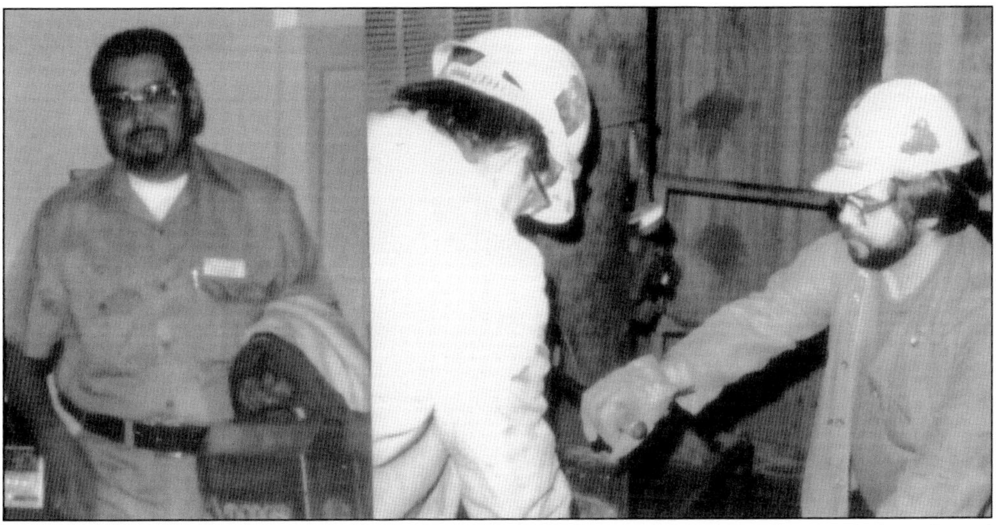

Julian Mendoza is pictured here ready for work with a special lunch bucket made by a friend who guaranteed it would be durable, which it was. He worked in the electrical maintenance department for 21 years.

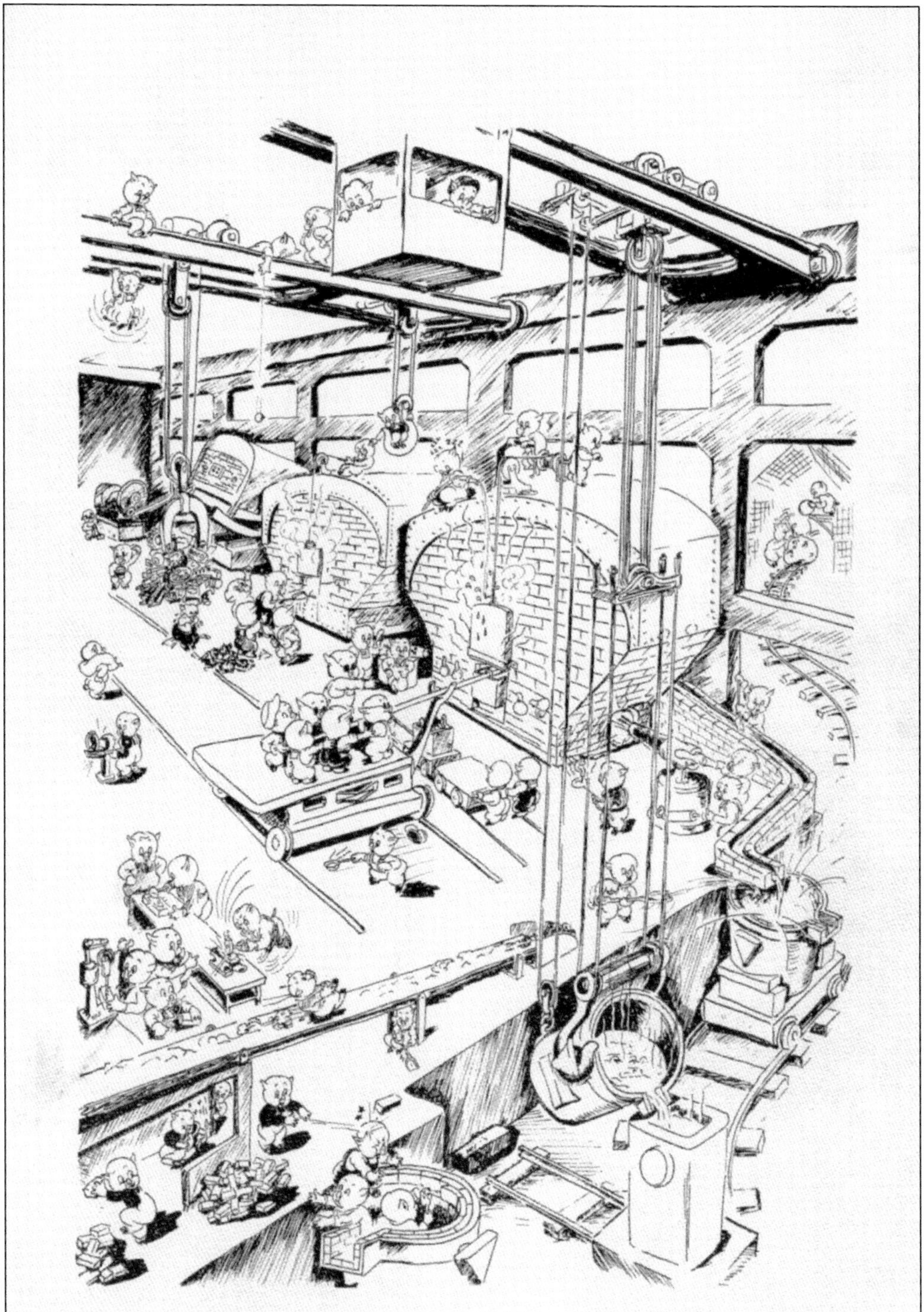

The piglets never ceased to be a symbol of Kaiser Steel Mill. Here they are depicted making steel for various products.

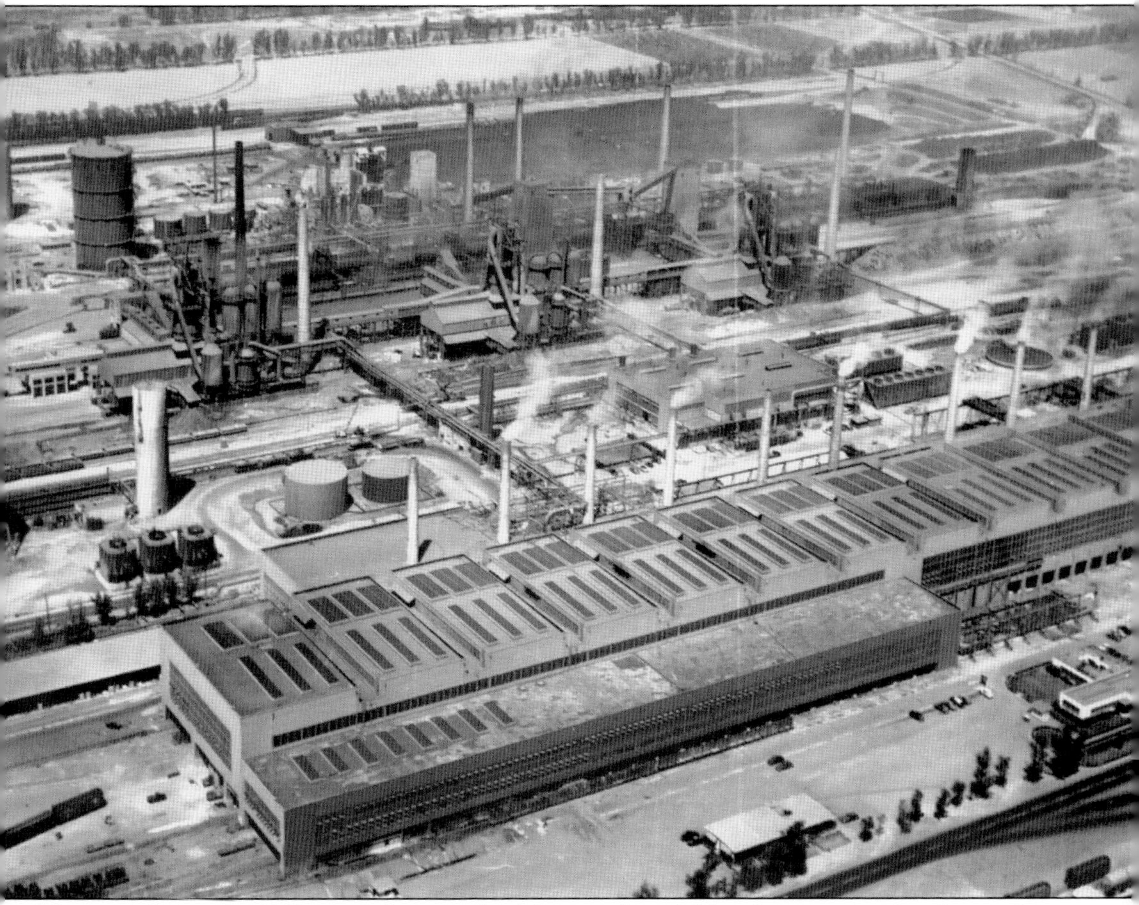

This is an August 21, 1953, aerial photograph of the open-hearth blast furnace area. Simply stated, steelmaking is a process of refining or purifying pig iron through controlled heat and chemical reaction. But in actual practice it is much more complicated. Kaiser Steel Mill operated six open-hearth furnaces housed in a structure 500 by 700 feet. In full operation, these furnaces produced 675,000 tons of steel ingots annually. These furnaces were like great bathtubs around and over which is a heat-resisting brick enclosure. At each end of the furnace was a regenerator, or vertical flue, filled with open brickwork that recovered some waste heat. Above the regenerators were two burners that provided the heat by burning a mixture of coke oven gas and fuel oil. The ingredients for steel—pig iron, commercial scrap, limestone, and iron ore—were charged through five doors in each furnace. One of the burners was then fired, directing its flame downward into the furnace, with the escaping heat collected in the regenerator on the opposite side. After a short interval, the first burner was turned off and the second on the other side was lighted, this time throwing the flame from the opposite direction over the charge and at the same time drawing out heat that had been collecting in the regenerator, thus providing more intense heat. This process was repeated over and over until a temperature as high as 3,000 degrees had been attained, creating a physical and chemical change in the charge. The result was a pool of molten steel covered with slag in the furnace. (Courtesy of Jonnie Long.)

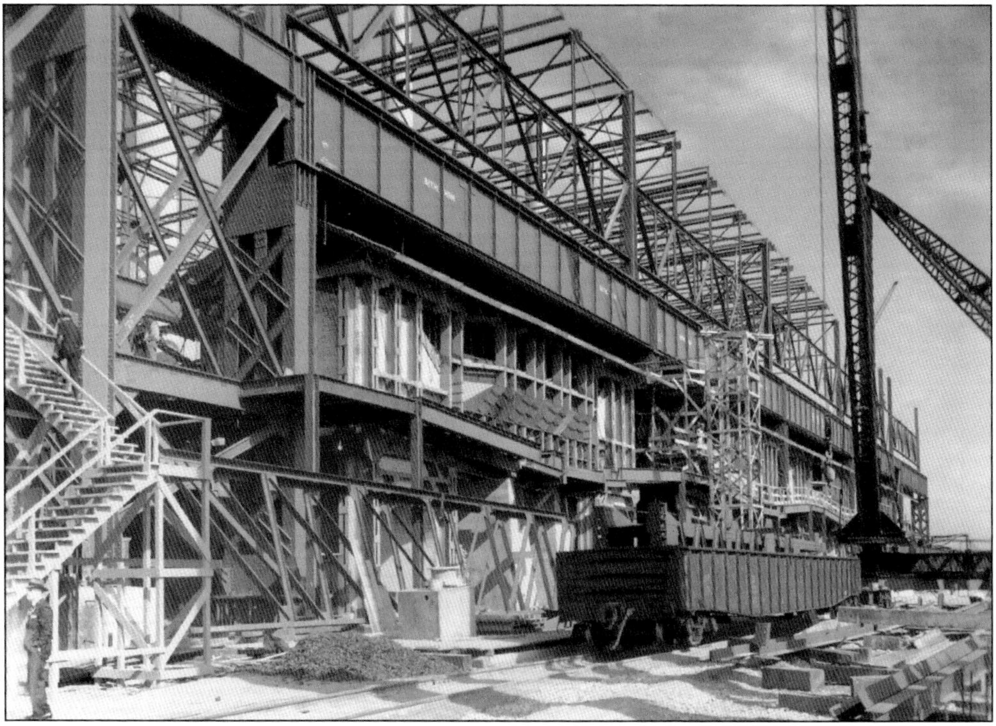

This is the open-hearth furnace looking northeast, on February 3, 1943. When the molten mass had been refined to the proper degree, the molten metal was tapped into a 200-ton, brick-lined ladle from which the white-hot steel was poured into ingot molds to solidify. (Courtesy of Alexia King.)

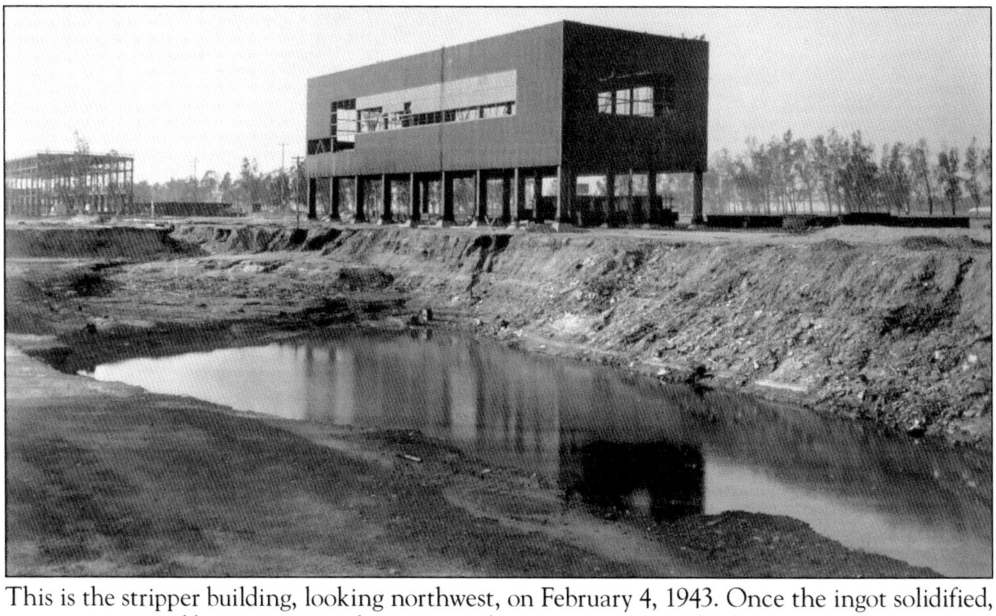

This is the stripper building, looking northwest, on February 4, 1943. Once the ingot solidified, it was transported by an ingot car, known as a "buggy," to a stripper building where the mold was removed, leaving nothing but the ingot of steel. These ingots were either sold on the commercial market or sent on to the rolling mill where it would be rolled into plate, structural steel or bars. (Courtesy of Alexia King.)

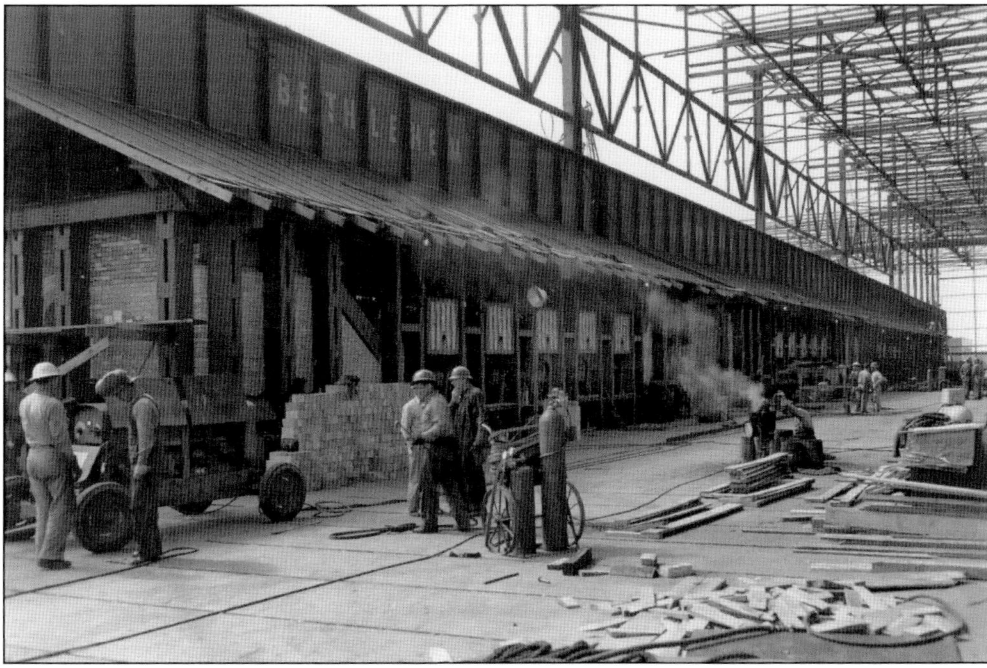

Open-hearth furnaces Nos. 1 through 4 are pictured looking west on the charging floor. (Courtesy of Alexia King.)

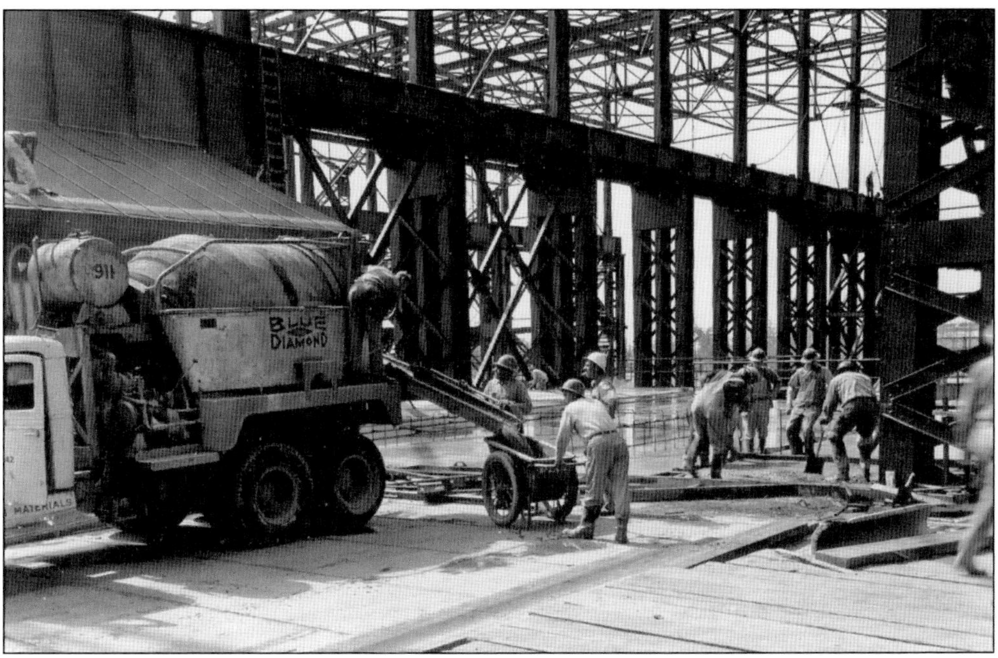

This is the open-hearth furnace building as concrete is being poured for the charging-floor slab. This photograph was taken on February 27, 1943. (Courtesy of Alexia King.)

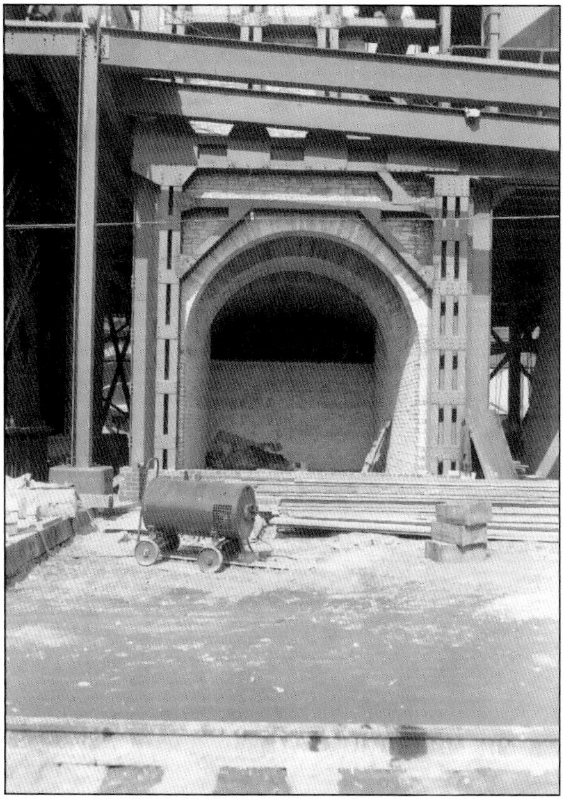

Open-hearth furnace No. 3 is depicted here on February 3, 1943, showing the brickwork on the furnace flue. (Courtesy of Alexia King.)

Firebrick is being installed here in open-hearth furnace No. 2. (Courtesy Alexia King.)

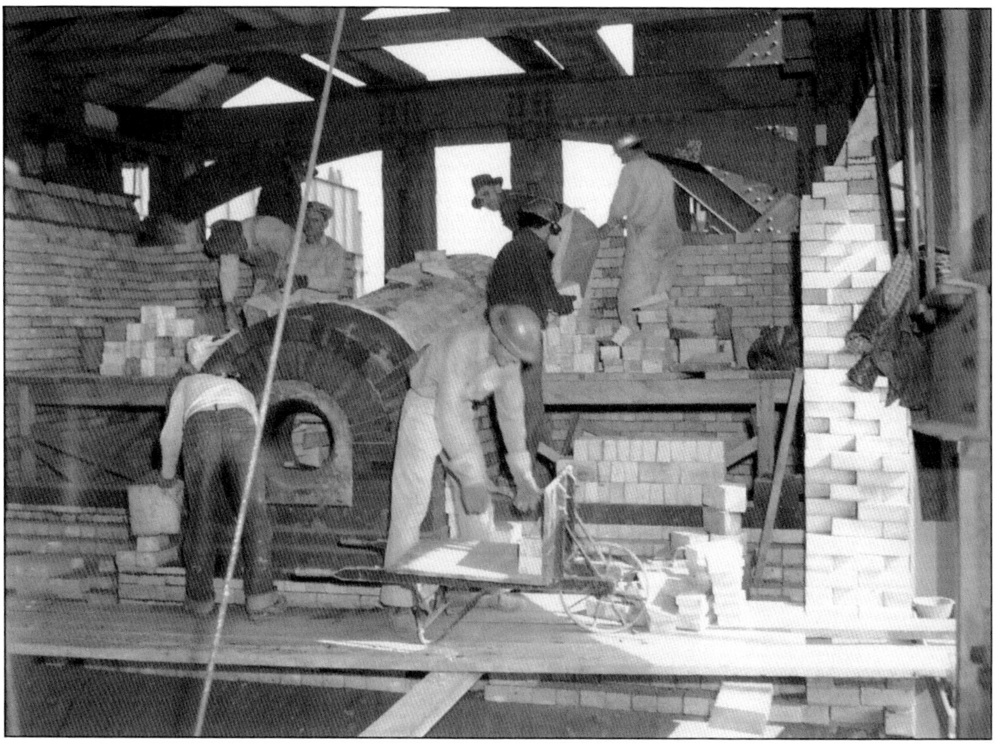

William Cajatan Blazer, who wore badge number 197, worked for Kaiser for 33 years as a brickman. He came from Homestead, Pennsylvania. His daughter, Cathy Elliott, and her husband, Grant Elliott, and Blazer's grandchildren, Lisa and John Elliott, work for California Steel. Other family members also worked at Kaiser or California Steel. Blazer was also, for 25 years, the financial secretary for Local No. 2869 of the United Steel Workers of America. (Courtesy of the Blazer family.)

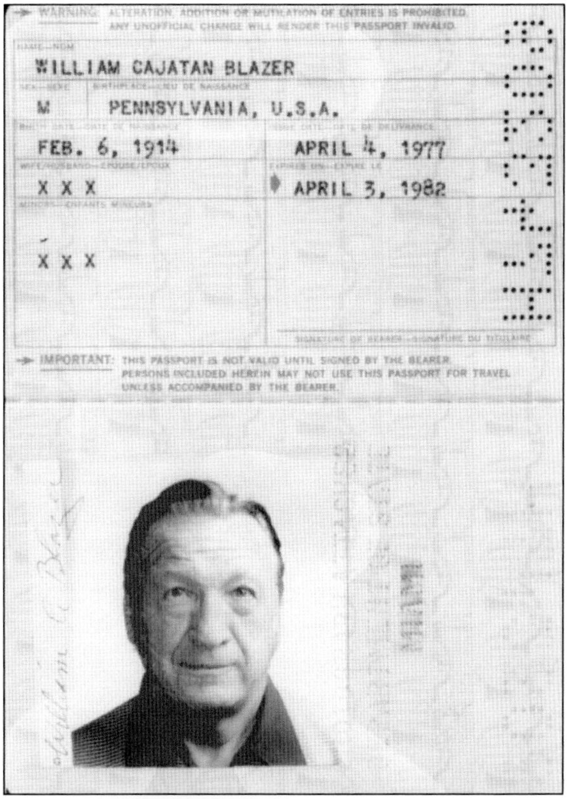

WARNING: ALTERATION, ADDITION OR MUTILATION OF ENTRIES IS PROHIBITED. ANY UNOFFICIAL CHANGE WILL RENDER THIS PASSPORT INVALID.

WILLIAM CAJATAN BLAZER

M

PENNSYLVANIA, U.S.A.

FEB. 6, 1914

APRIL 4, 1977

X X X

APRIL 3, 1982

X X X

IMPORTANT: THIS PASSPORT IS NOT VALID UNTIL SIGNED BY THE BEARER. PERSONS INCLUDED HEREIN MAY NOT USE THIS PASSPORT FOR TRAVEL UNLESS ACCOMPANIED BY THE BEARER.

This image, taken February 3, 1943, shows the interior of open-hearth furnace No. 1.

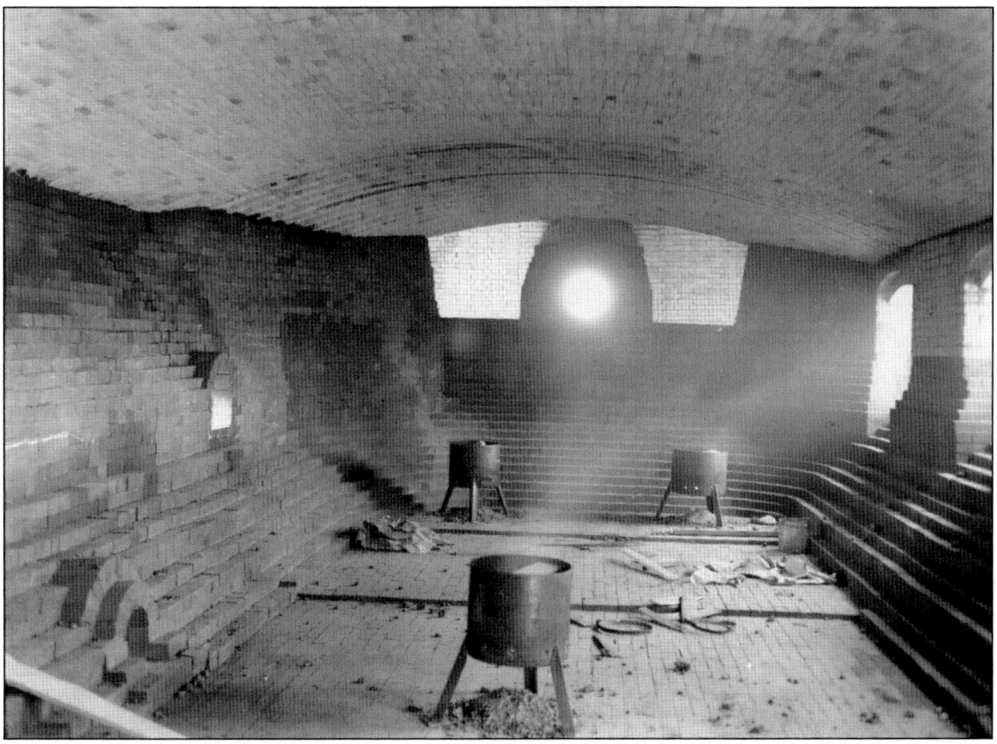

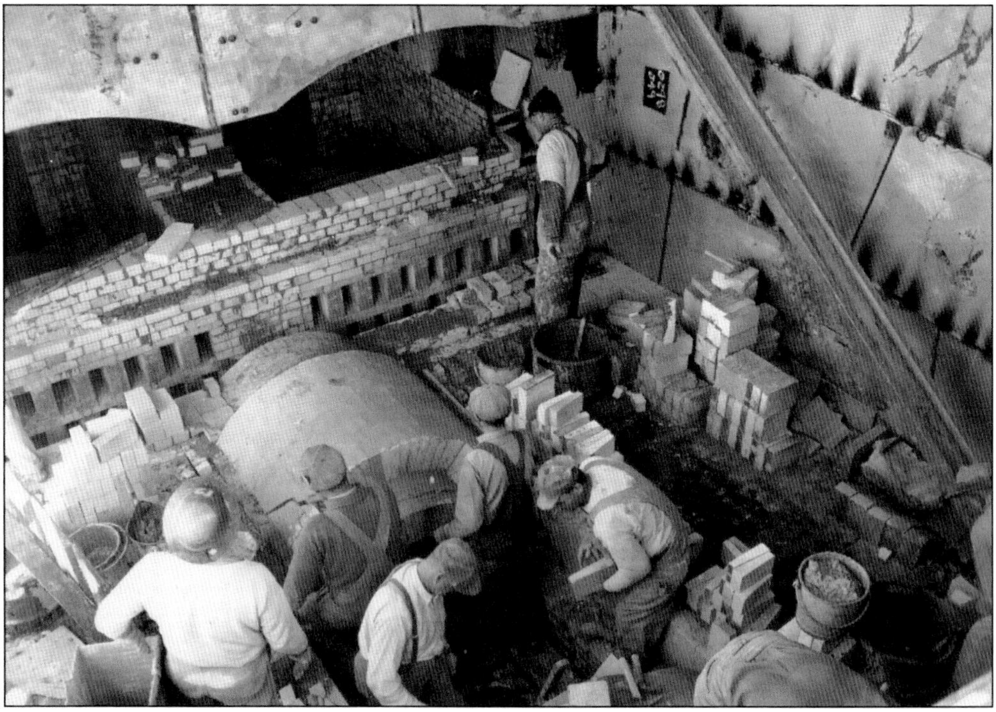

The interior of the furnace brickwork of open-hearth furnace No. 1 is pictured here. (Courtesy of Alexia King.)

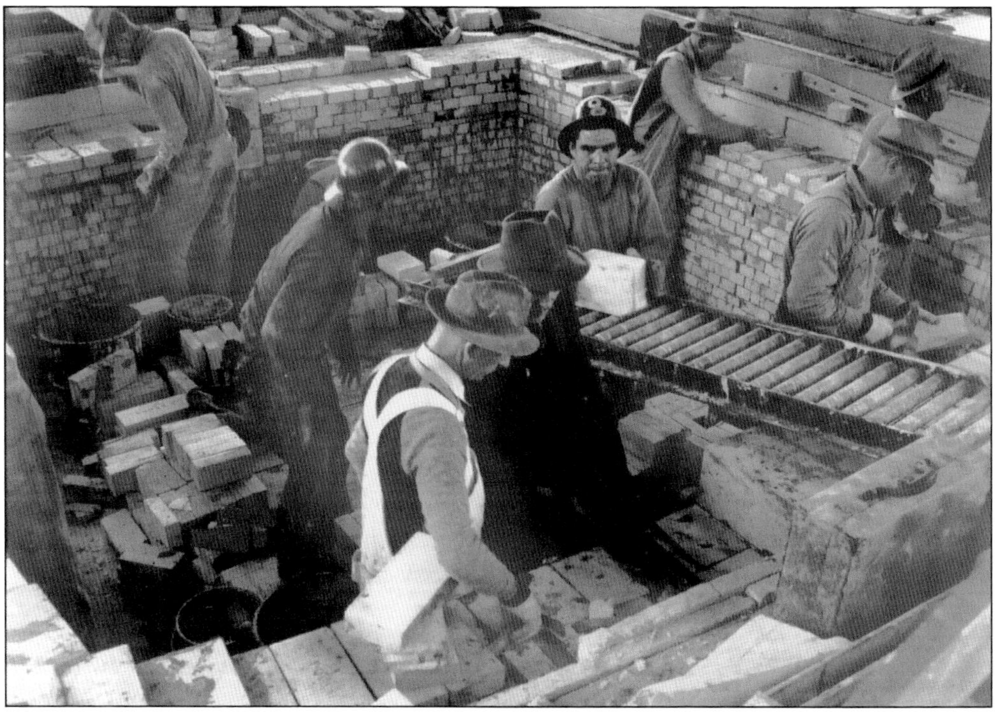

This February 3, 1943, photograph shows the installation of firebrick in the soaking pit. (Courtesy of Alexia King.)

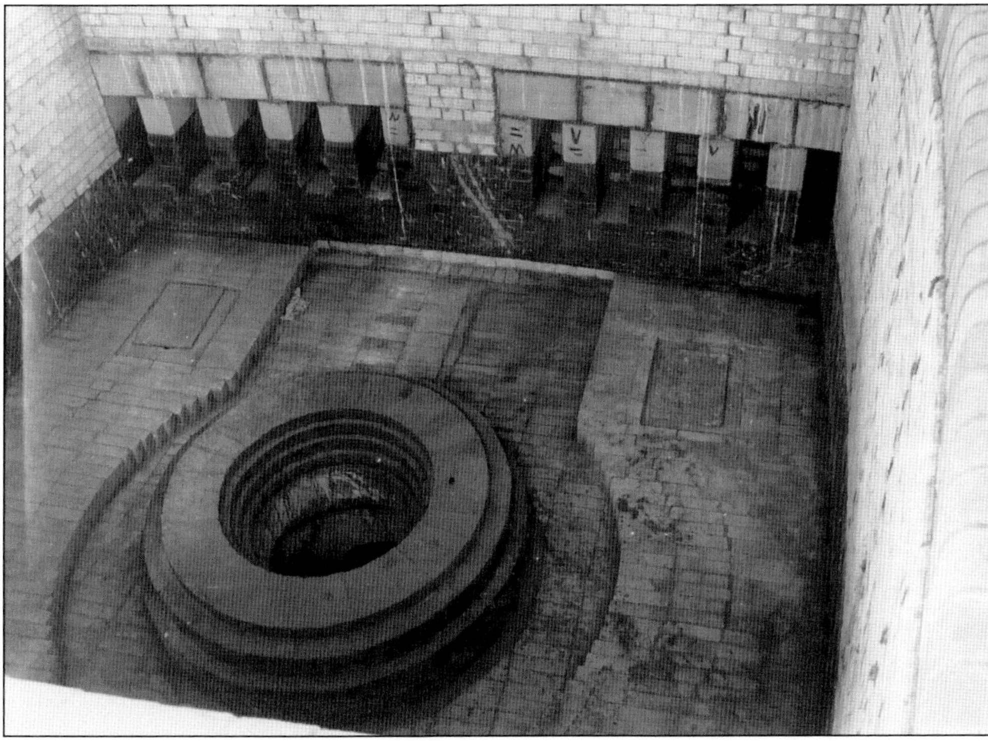

The soaking pit is pictured here, completed, on February 27, 1943.

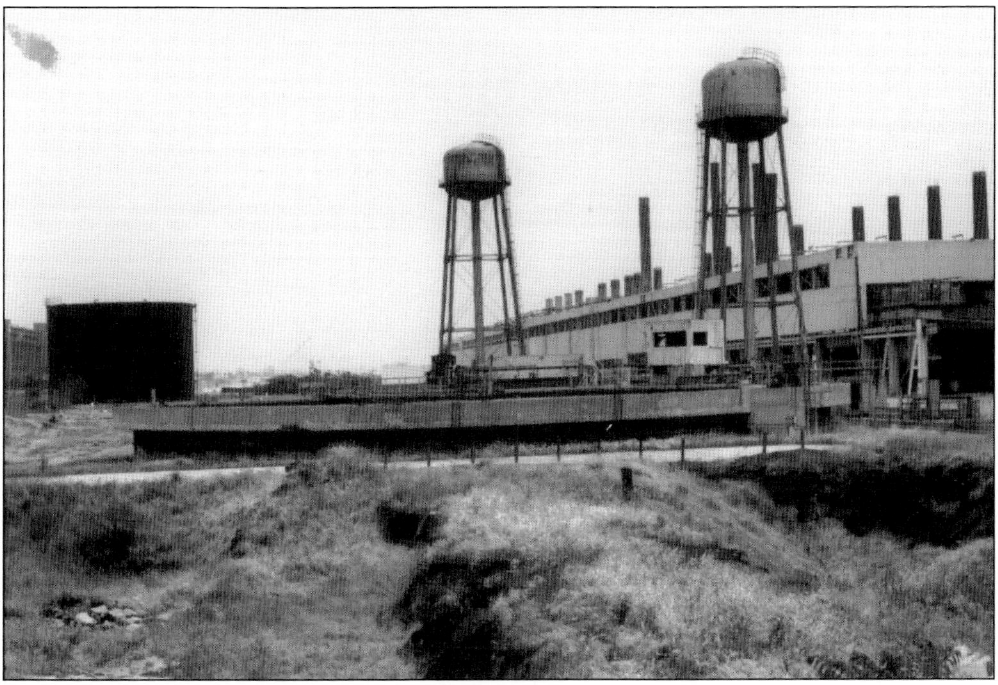

These are the water towers on the north side of the soaking pit. Both towers have since been taken down, but the metal dedication plates were saved for the Fontana Heritage Museum Association, Kaiser Museum Section. (Courtesy of Charles Dall.)

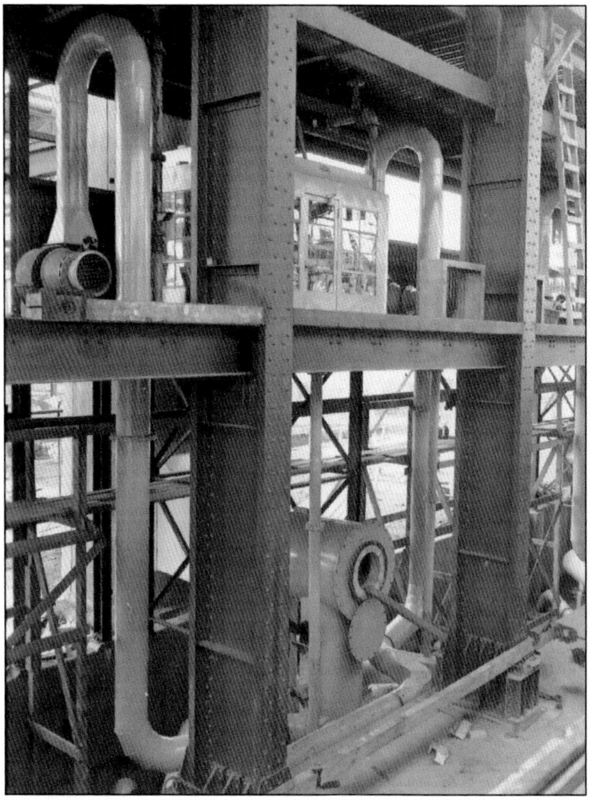

The enclosure here is the soaking pits control house, which was known as Battery No. 3.

The southwest end of the soaking pits is pictured here in June 2005.

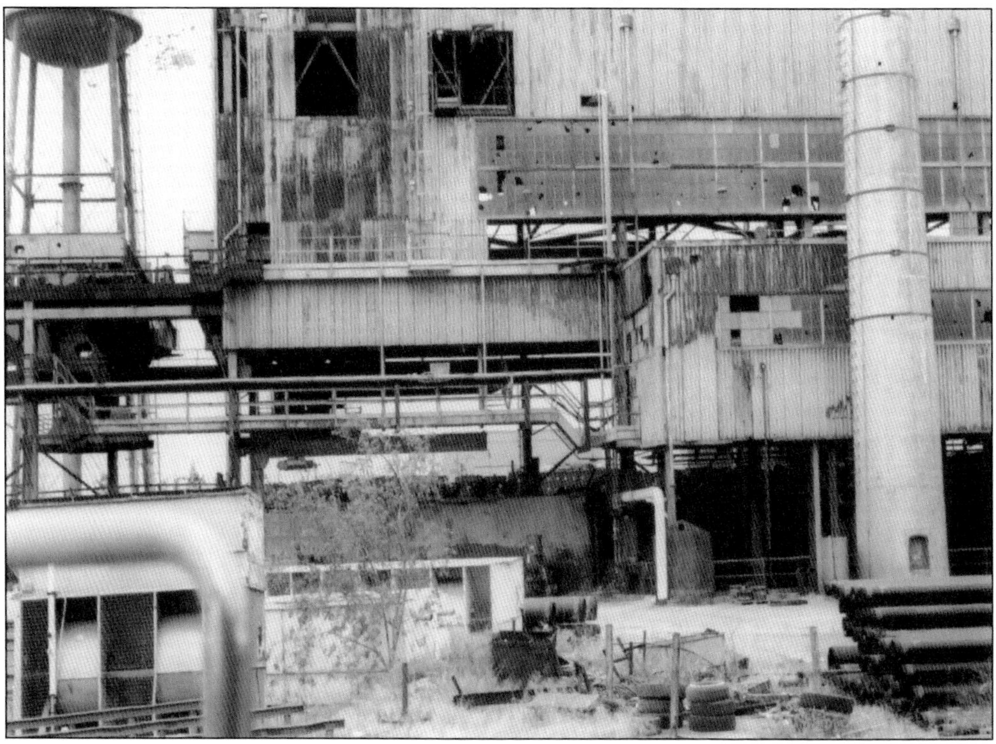

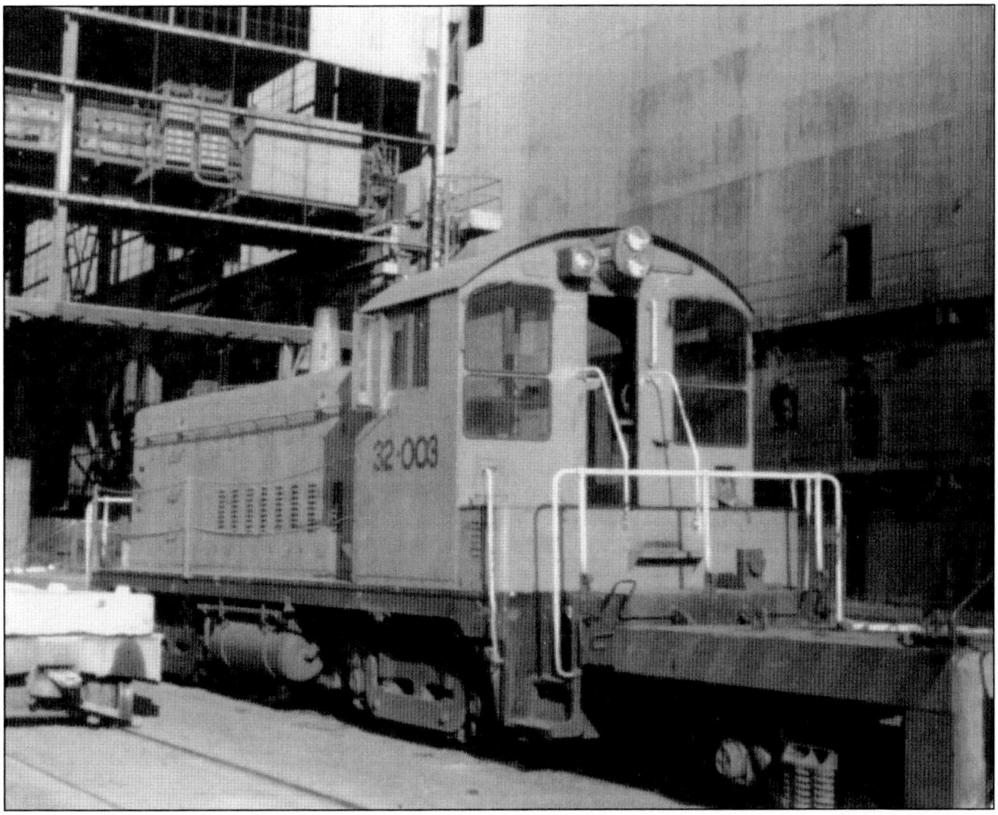

This engine at California Steel was used at Kaiser Steel and is still on the site. This photograph was taken in May 1998.

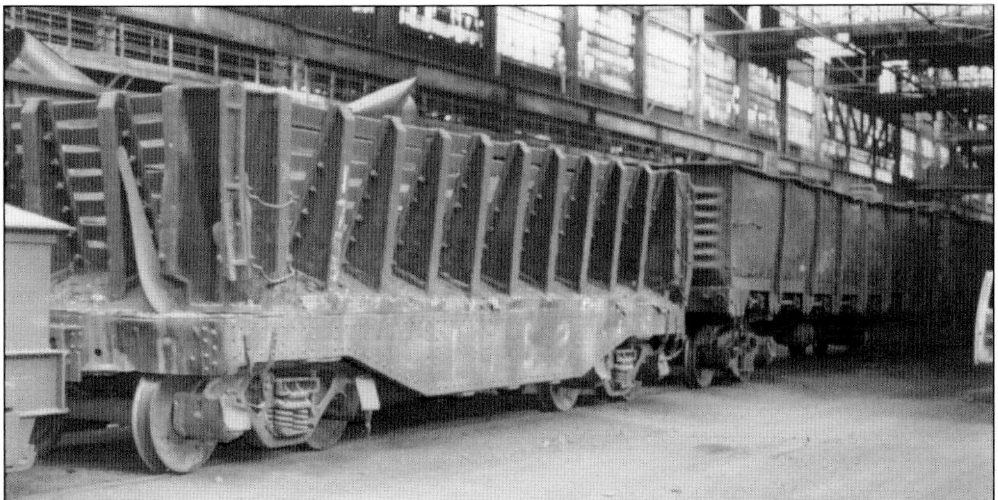
This "cattle car" dates back to around 1929. It was used in the movie *The Outlaws* and was eventually cut up for scrap. The building was once the structural mill and is now the California Steel scrap yard.

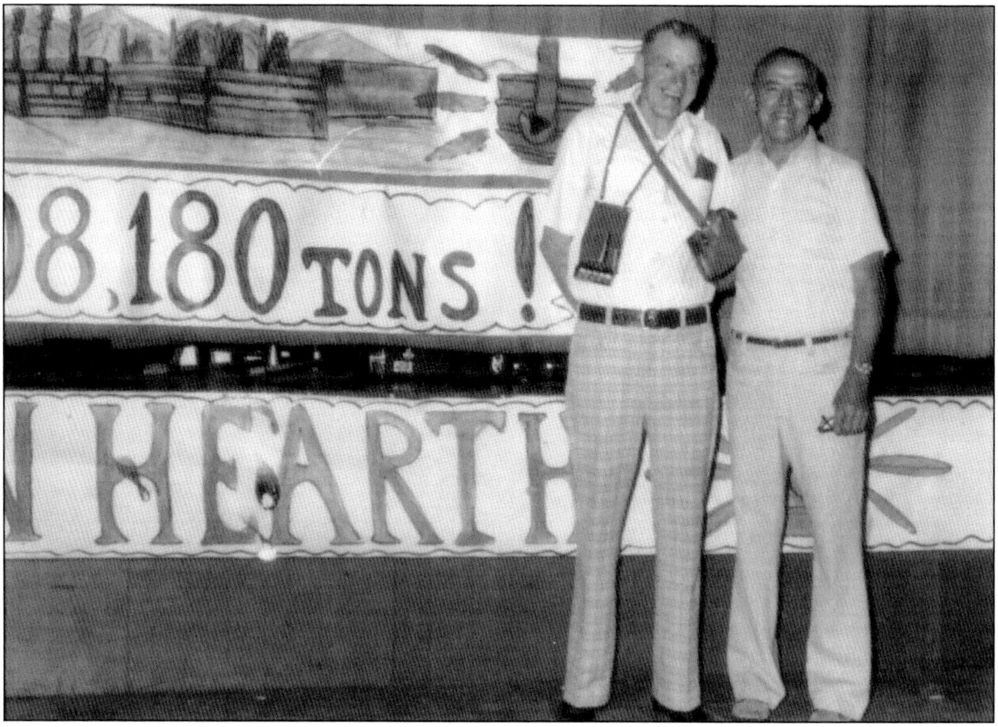

Two brothers, Joseph Scanlon and Walter Scanlon, pictured here, both worked in the open hearth for approximately 33.5 years. They were from Barberton, Ohio.

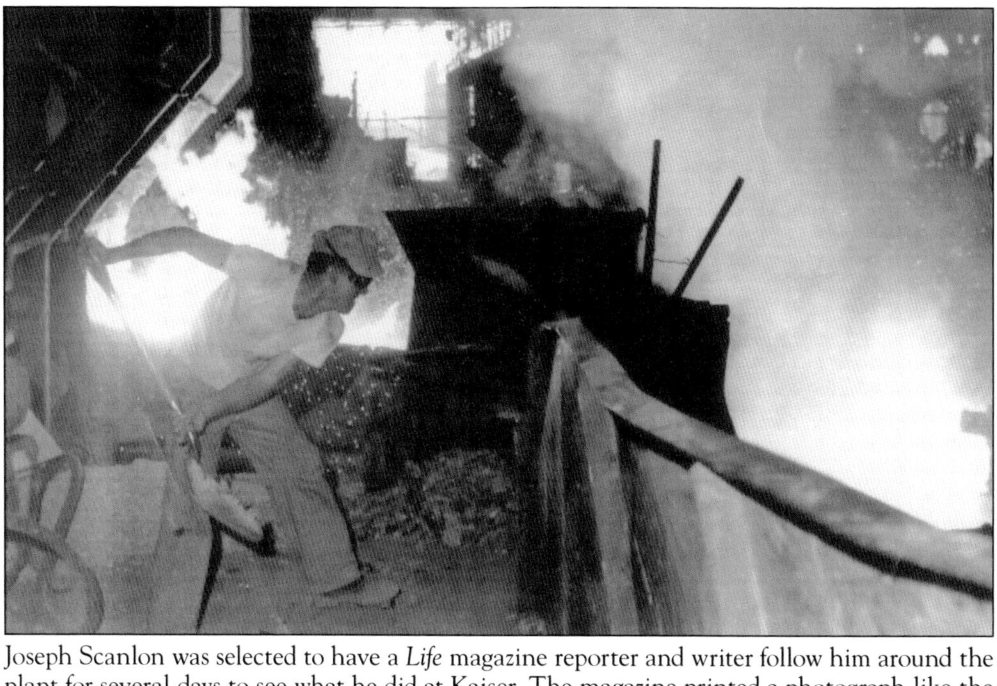
Joseph Scanlon was selected to have a *Life* magazine reporter and writer follow him around the plant for several days to see what he did at Kaiser. The magazine printed a photograph like the one here, showing him at work. This was done during the 1959 strike. The magazine sent him a letter apologizing for not putting more than one image in the issue.

California governor Goodwin J. Knight pauses during his whirlwind tour of the Kaiser Fontana Steel Mill on a Wednesday afternoon. He stopped for a friendly photograph with open-hearth workers as molten metal blazes in the background. Pictured with him, from left to right, are second helper Joe Scanlon, third helper K. Jenson, Moorad Alder, and melter Chester Crane Jr.

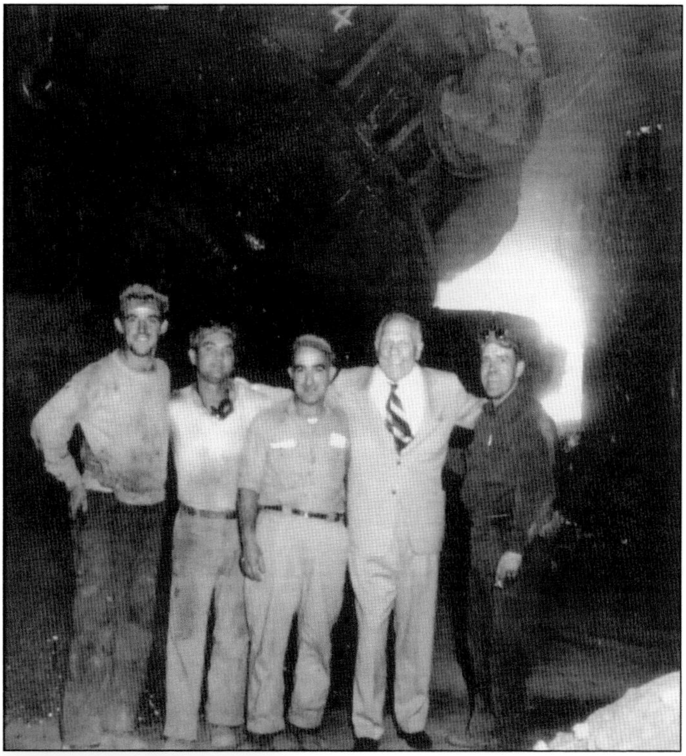

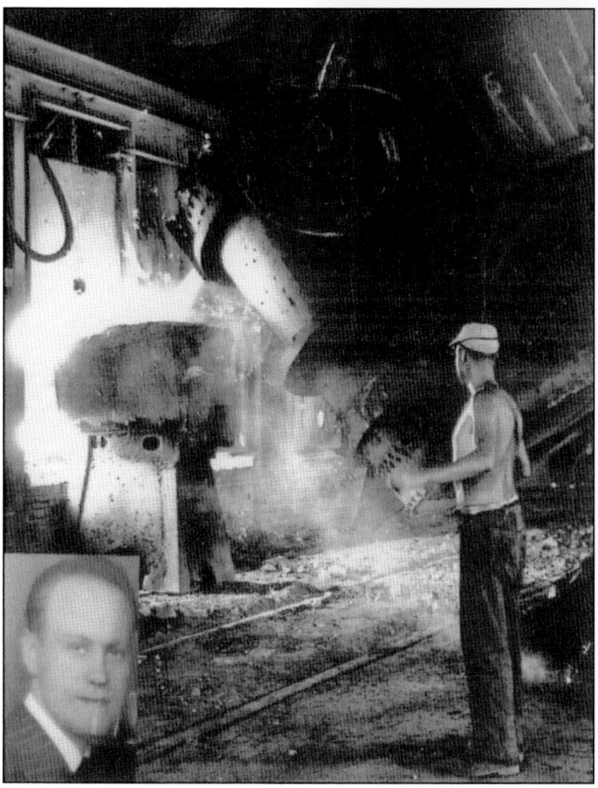

Rudolph "Rudy" Pock, originally from Bethlehem, Pennsylvania, worked for 11 years before opening a music store in Ontario. In the open-hearth building, he's called a "first helper," which is a rather modest title to pin on a man who assembles, mixes, and cooks the ingredients for 225 tons of high-grade steel. Pock was first helper on the No. 7 open-hearth furnace. This photograph appeared in *Westward* magazine. (Courtesy of Mrs. Pock.)

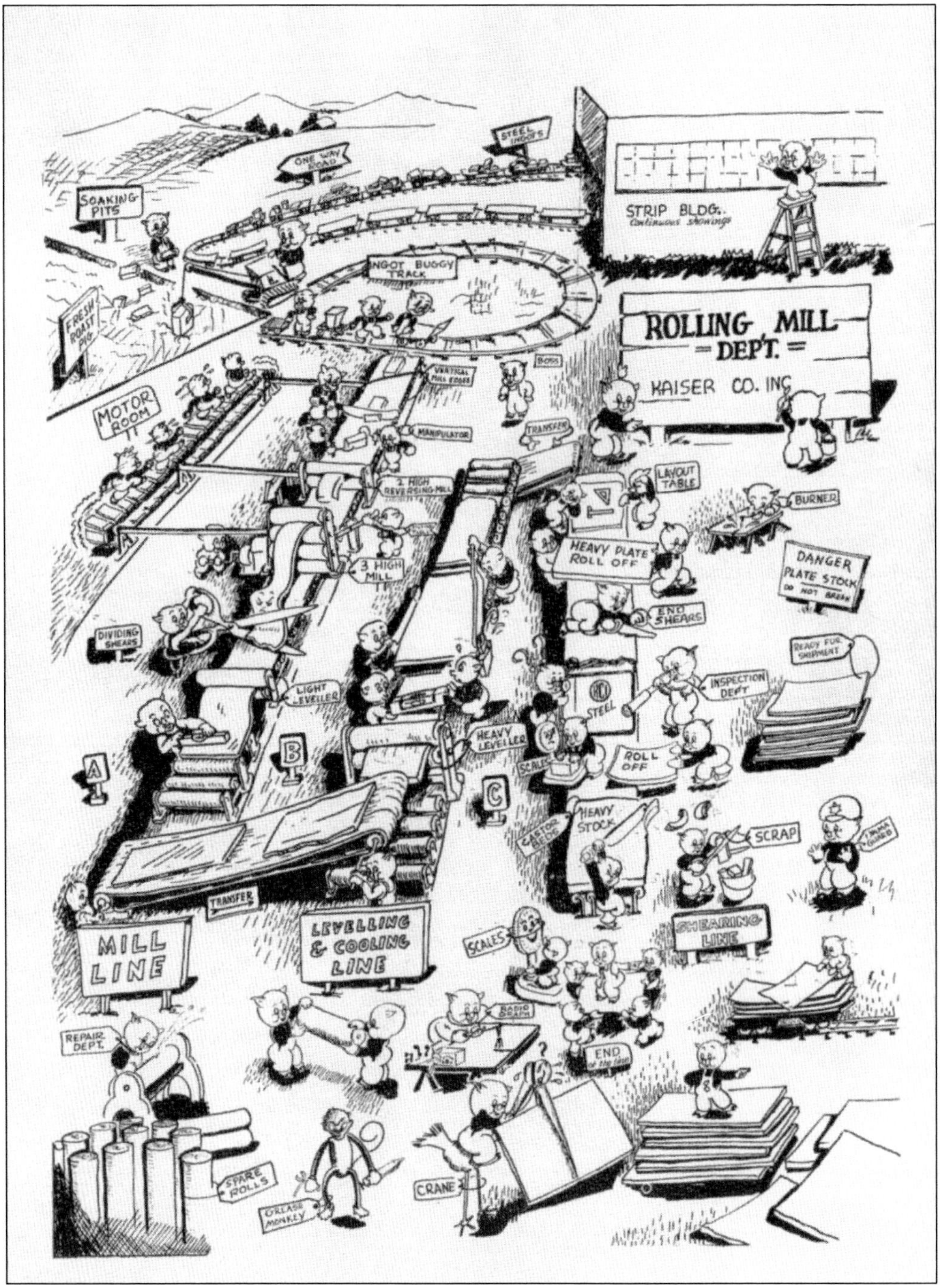

Piglets are pictured in the work areas of the rolling mills. Ships sailed, buildings towered, planes flew, and automobiles traveled by using steel rolled at Kaiser Company. The rolling mills are the largest structures at the steel mill, covering an area of 683,000 square feet. Here they carry on the process of making steel with the open hearth left off, taking the ingots from the stripper building to the soaking pits at the north end of the structure. In these soaking pits, ingots are heated to a uniform 2,250 degrees so they become sufficiently malleable for rolling.

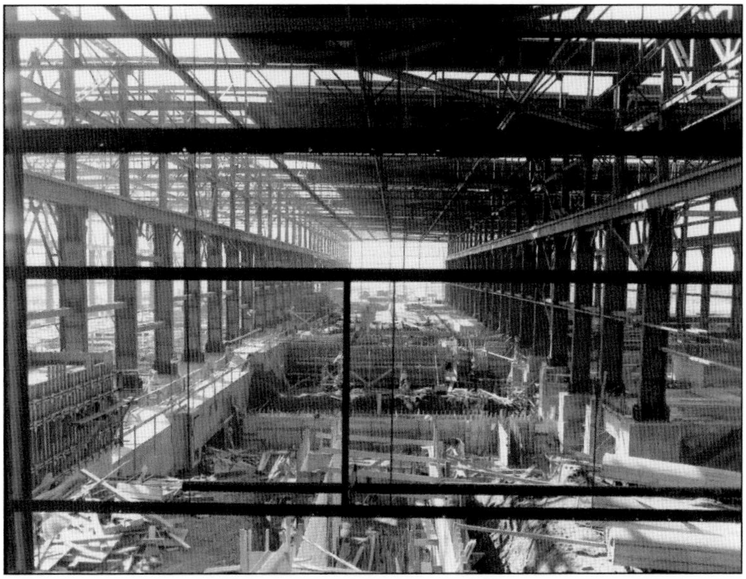

The construction of the plate-mill building, looking south over the power basement, is pictured here on February 27, 1943. (Courtesy of Alexia King.)

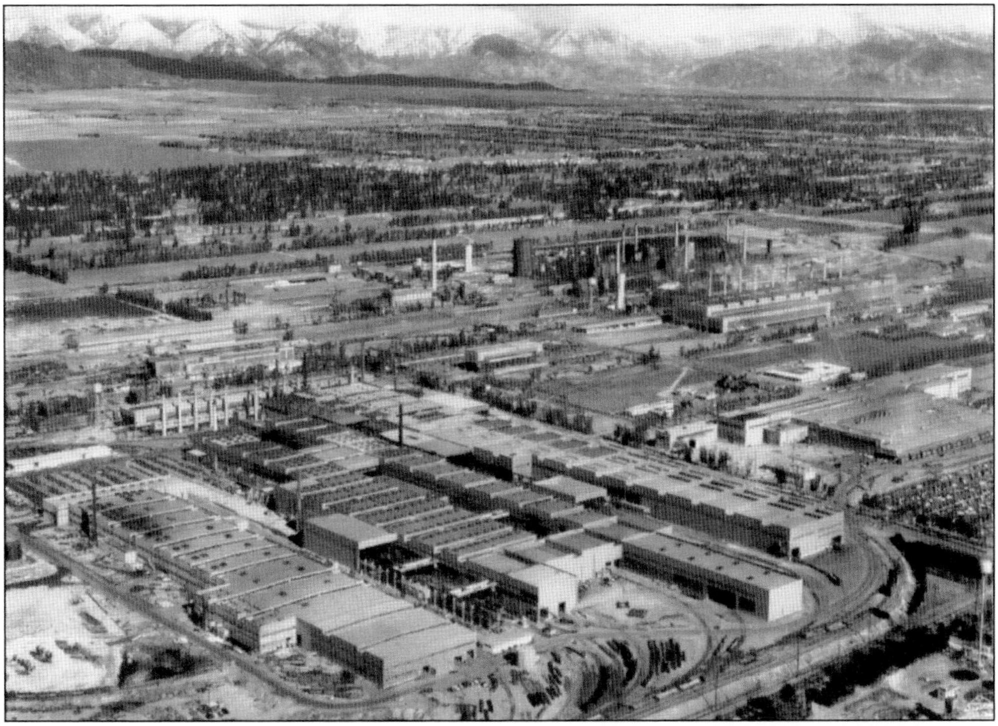

This aerial view takes in the southwest section of Kaiser, showing the plate, rolling, finishing, and hot-strip mills. Plate mills may be compared to huge washing-machine wringers, through which the ingots are squeezed and flattened to the required thickness, width, and length. These mills are able to take a single ingot and roll it until it becomes a plate of tough steel that is 110 inches wide and 50 feet long. From this are sheared plates of any desired length and width. In the plate mill, ingots are first rolled in a roughing mill, which consists of two large steel rollers through which the ingot passes back and forth as the rollers are brought closer and closer together, forming a rough, thick plate. (Courtesy of Kaiser High School Library.)

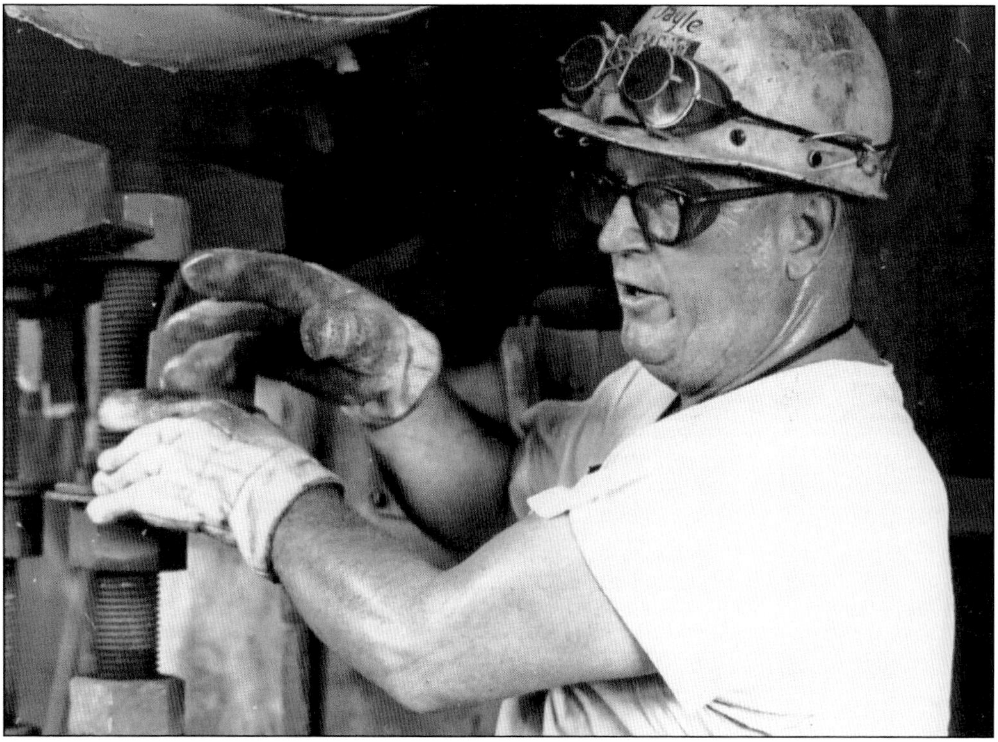

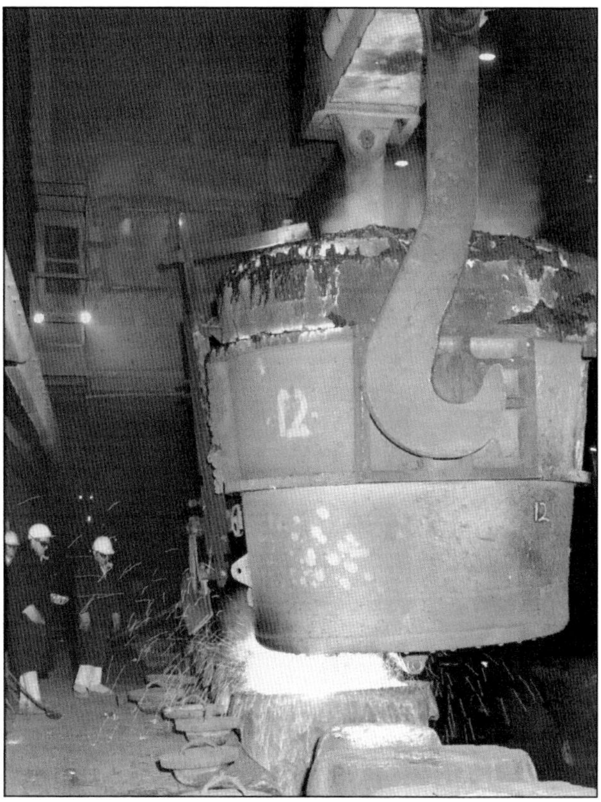

Dale Pipler, who worked as the first casting man at Kaiser, is pictured here. He worked at the plant for 33 years after arriving from Nebraska.

This is a view of the area where Dale Pipler worked.

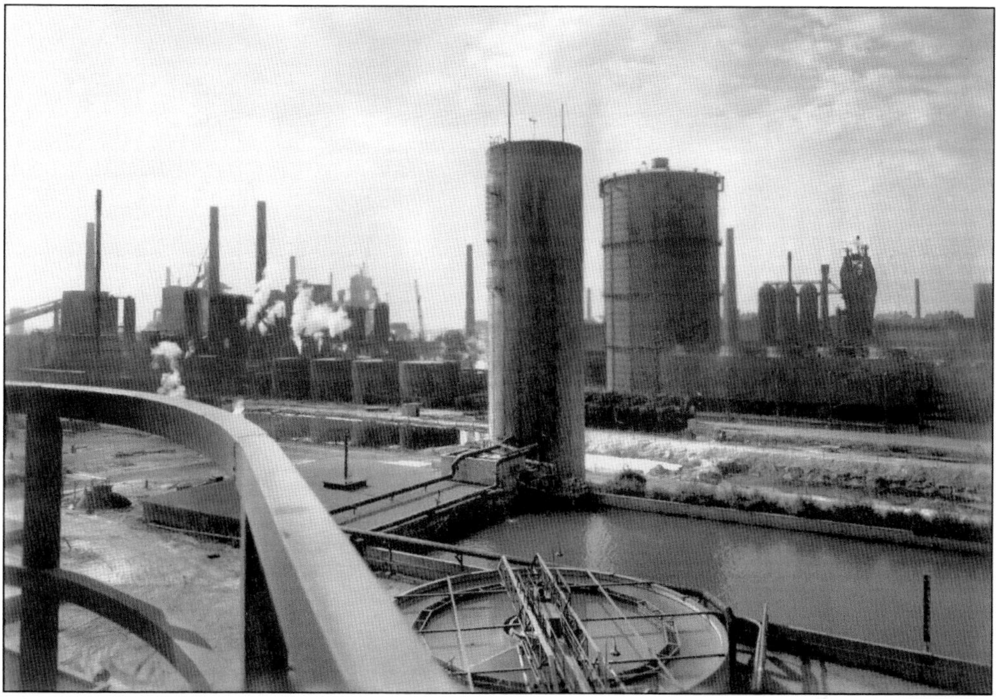

The anemometer, which measures wind force and velocity, is pictured with other apparatus on August 11, 1958. (Courtesy of Charles Dall.)

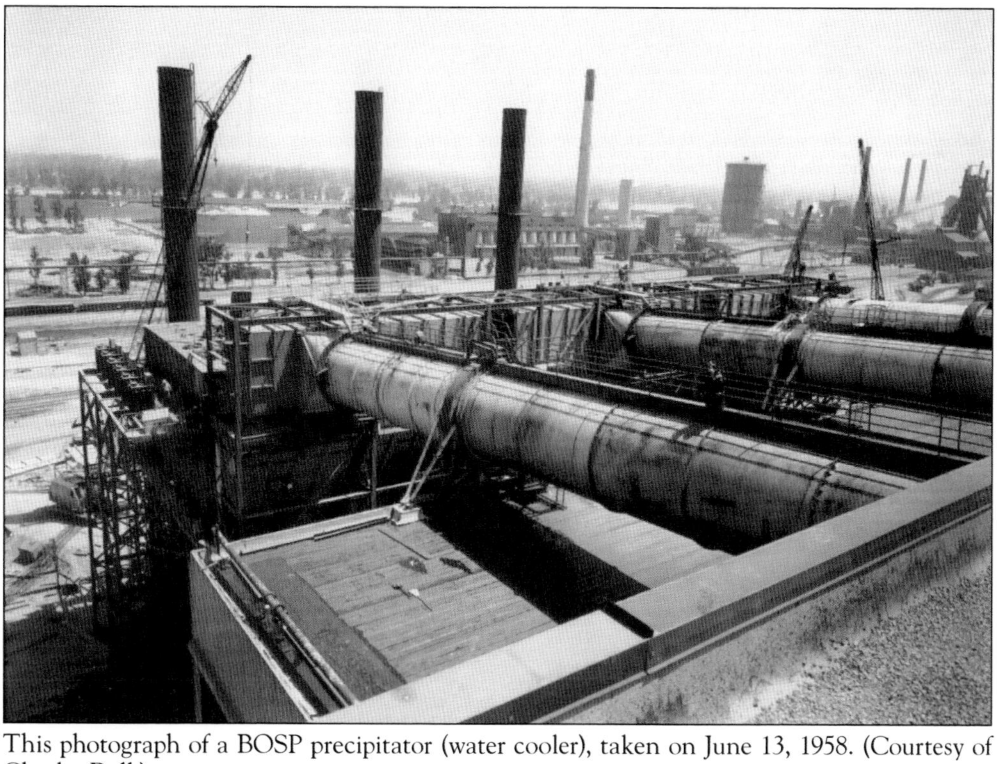

This photograph of a BOSP precipitator (water cooler), taken on June 13, 1958. (Courtesy of Charles Dall.)

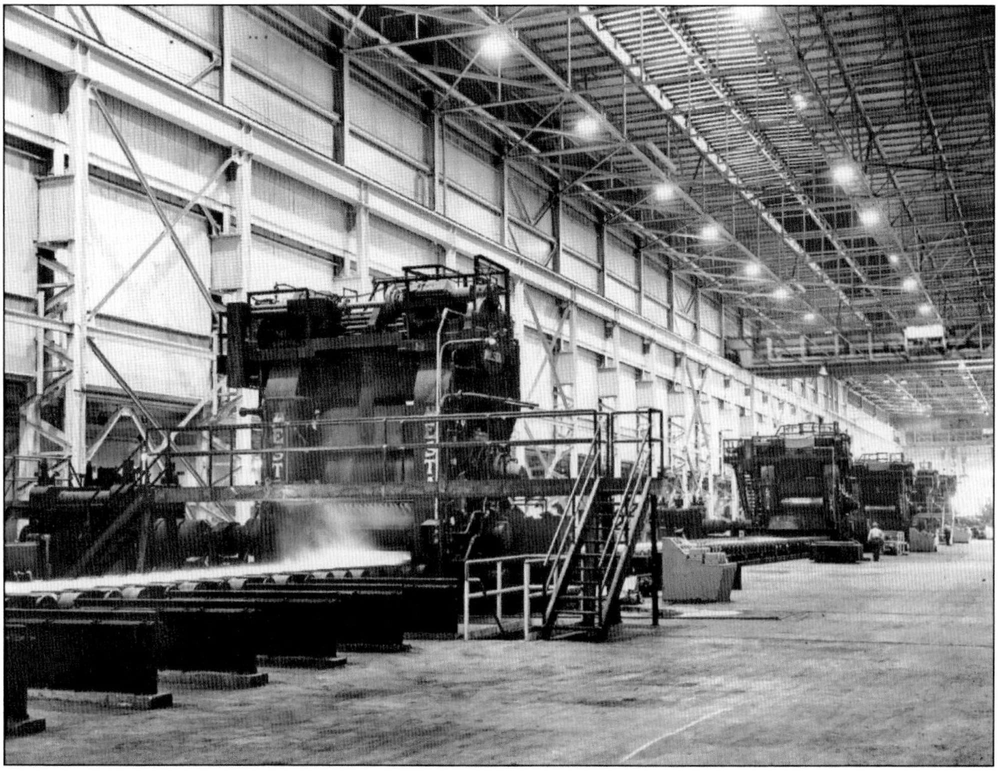

Roofer R5 enters into the finishing mill around 1960. (Courtesy of Bill Boatman.)

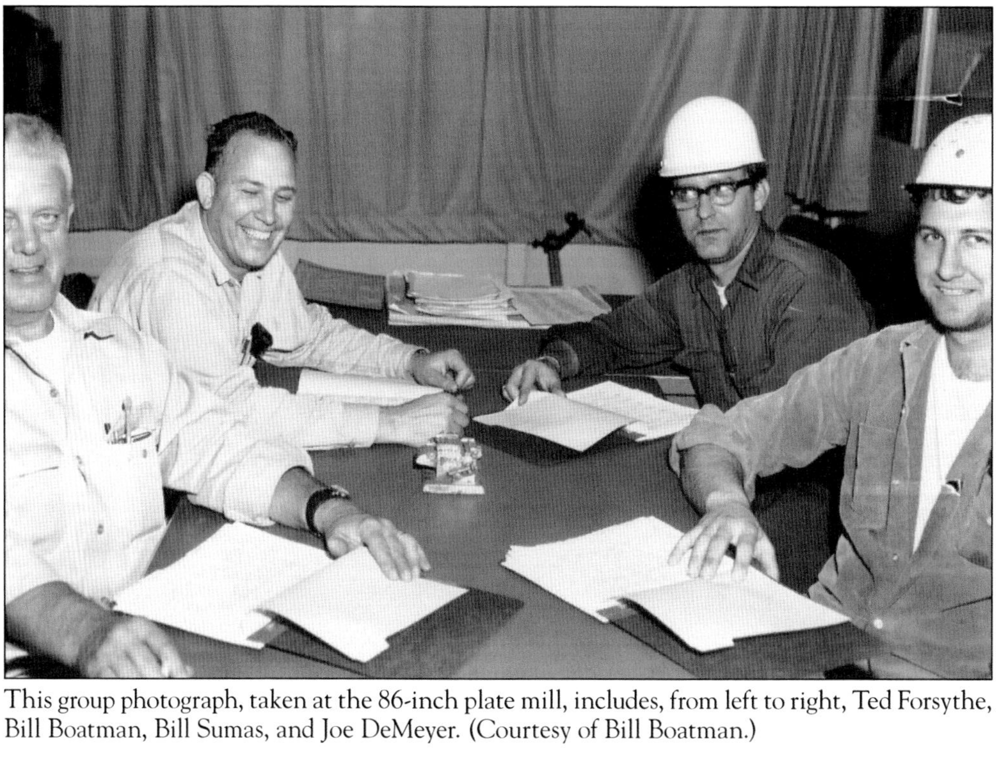

This group photograph, taken at the 86-inch plate mill, includes, from left to right, Ted Forsythe, Bill Boatman, Bill Sumas, and Joe DeMeyer. (Courtesy of Bill Boatman.)

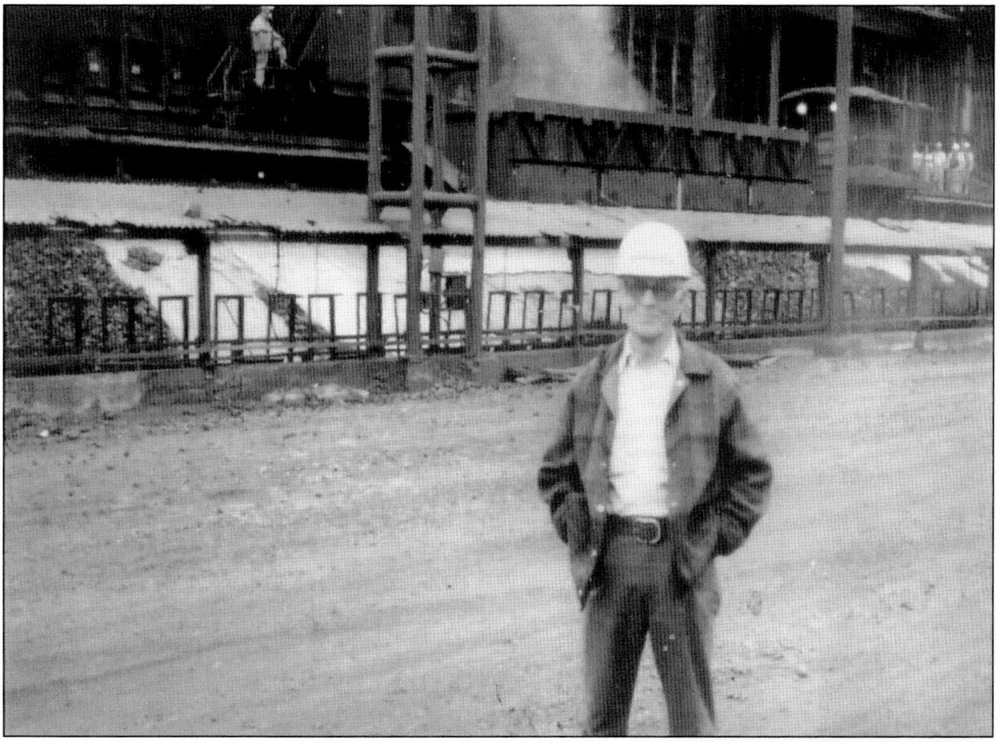

Les Price was foreman of the by-products on the last push of the coke ovens in December 1982.

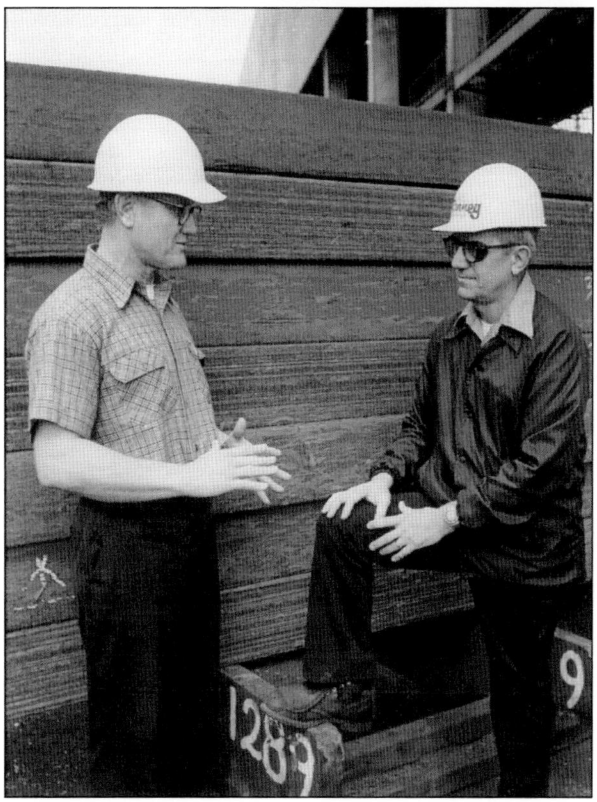

This is a photograph of Earl Ranney, the plate-mill foreman, who worked for 32 years with crane operator Mr. Stanfield at the slab yard/plate mill. Frank Carmody, who was a foreman of the hot-strip finishing mill, was a friend of Earl Ranney.

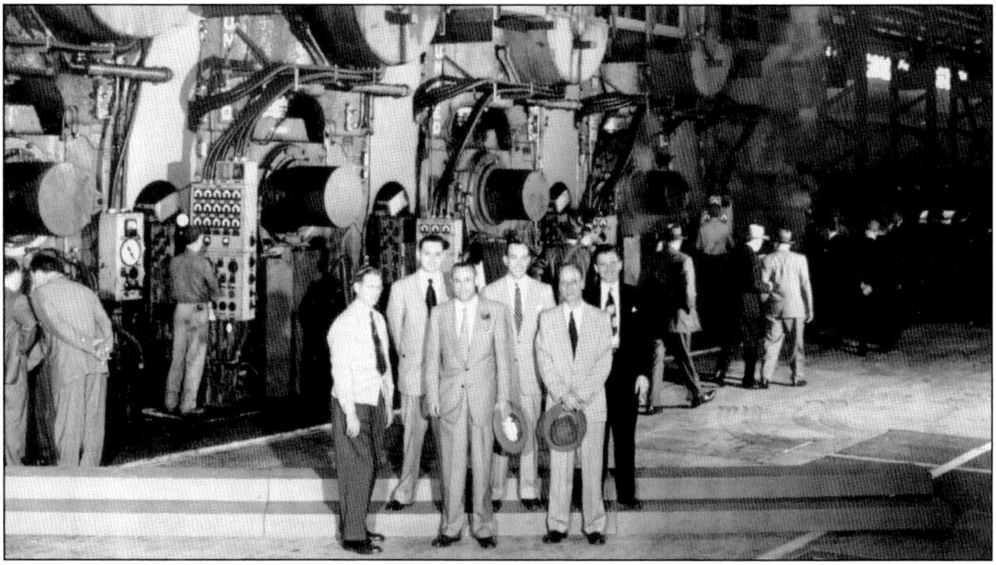

Al Lynch (first row, left), Mr. Reynolds (center), Button Seineke (first row, right), and Mr. Droessel (second row, right) conducted the inspection of the newly redone hot-strip mill. The first slab rolled onto the 86-inch hot-strip mill on February 15, 1950. (Courtesy of the McCarthy family.)

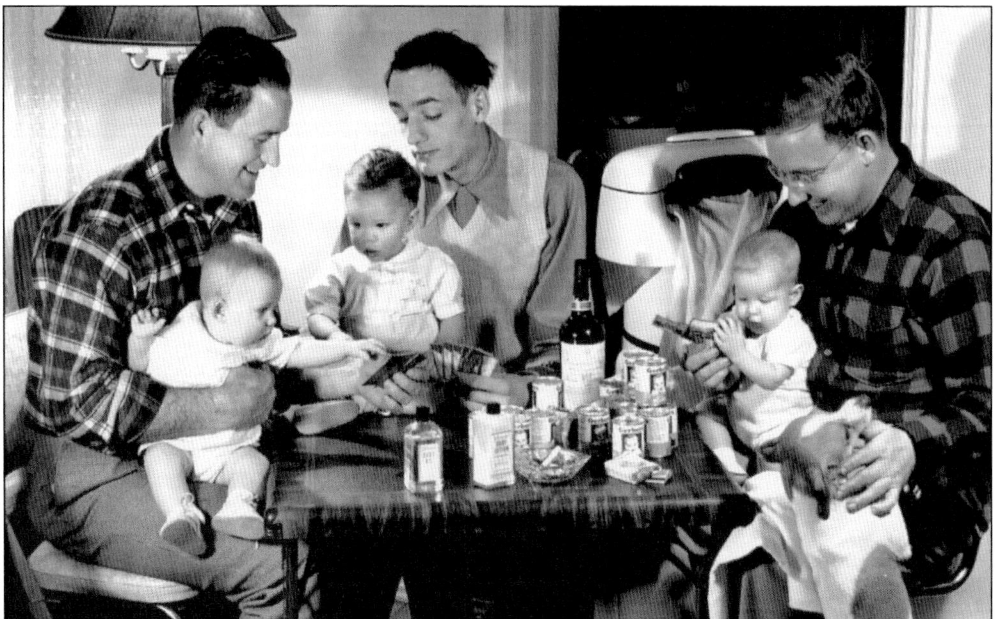

When moms head out for the evening, the dads have to babysit. These men all worked as cold strippers. The kids were begging for a game, and they got one—dealers-choice poker with canned baby food as chips. Pictured here, from left to right, are "Tex" Wallace, who gets some fancy finger-pointing from daughter Diana (6 months old) and some resentment from Edward Gilmore (10 months), who is overridden by super kibitzer Frank Gilmore (assistant roller); then comes the washing machine, which grinds out fresh, dry diapers in case of any "unusual" moisture at the table; and Charles Stanley Jr. (6 months) fans his bobtailed flush while Charles Sr. (roller) pats his dog Fifi. It all took place in the Stanley home at 183 South Bennett Street in Fontana.

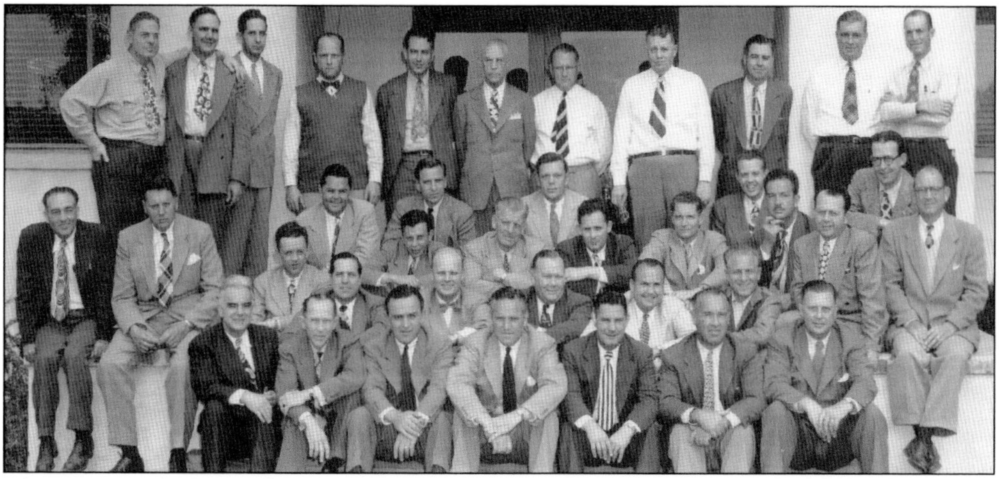

The Kaiser Steel sales conference in Fontana in June 1948 is memorialized in this group photograph. The participants are, from top to bottom, Depew, Jacobson, Howard, Pelletier, Madden, Ordway, Lenhart, Rich, Miller, Herbert, and Edward Alvin Gilmore, who was an executive for eight years, hailing from the steel hub of Pittsburgh, Pennsylvania. In the second row, from the top, are Macorey, Asquith, Gleason, Cooper, Reniers, Hole, and Jones. The third row includes Cosgrove, Johnson, Babylon, Etcheverry, Dunnigan, Hogan, Conley, and Ashby. The fourth row includes Duffy, Carlson, Balaam, Emerson, and Beyma. In the fifth row are Osborne, Lamborn, Ede, Borden, Gulliver, Griffith, and Hoagland.

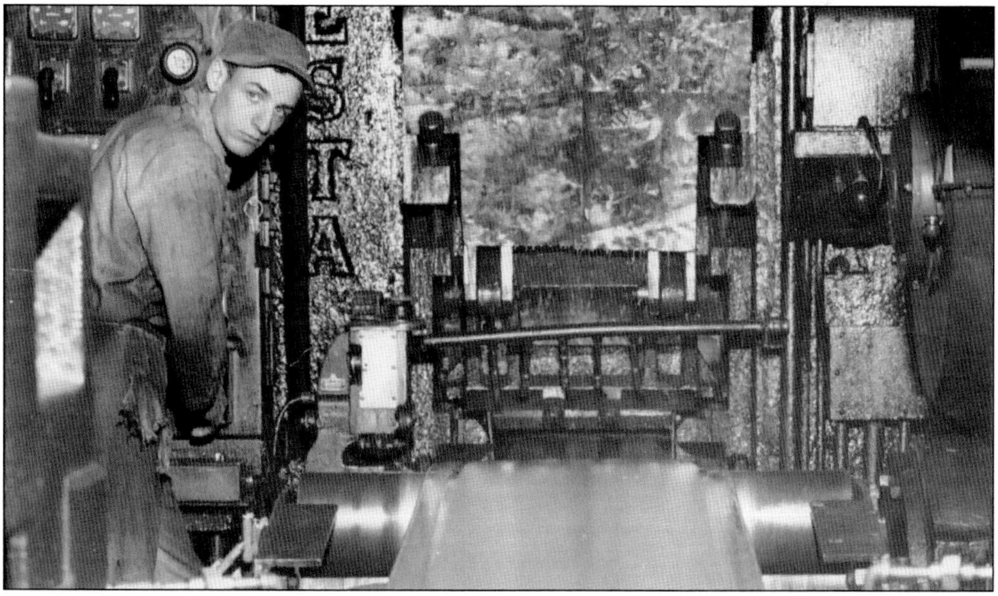

Frank Gilmore worked at the cold-roll mill for 25 years. He later became the foreman of the steel-galvanizing reversing mill. His father was Edward Alvin Gilmore, pictured here. From the roughing mill the plate goes to a finishing mill, where it is rolled still thinner and smoother. The rolls can be adjusted so that plates of any desired thickness could be made, ranging from a quarter-inch to two inches. The plate passes through a leveler, which removes any unevenness, after which Guillotine shears cut the ends, making plates of any desired length, and side shears cut plates that cannot be sheared on the rotary shears. After trimming, the plate is marked, sized and stamped, and sent to jaw shears that cut the ends, while side shears square the sides.

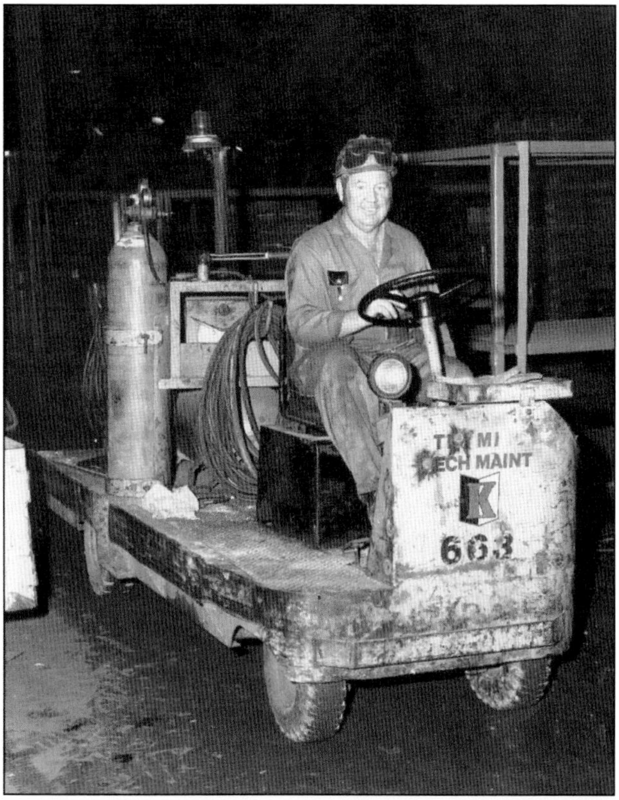

Jesse "Red" Williamson Jr. is pictured on his scooter, which he built at Kaiser, to get around the enormous plant to do the needed welding. A product of Georgia, he worked for 20-plus years at Kaiser.

Red Williamson does his job here of welding in the forge, pipe, and tin mills. (Courtesy of Margie Williamson.)

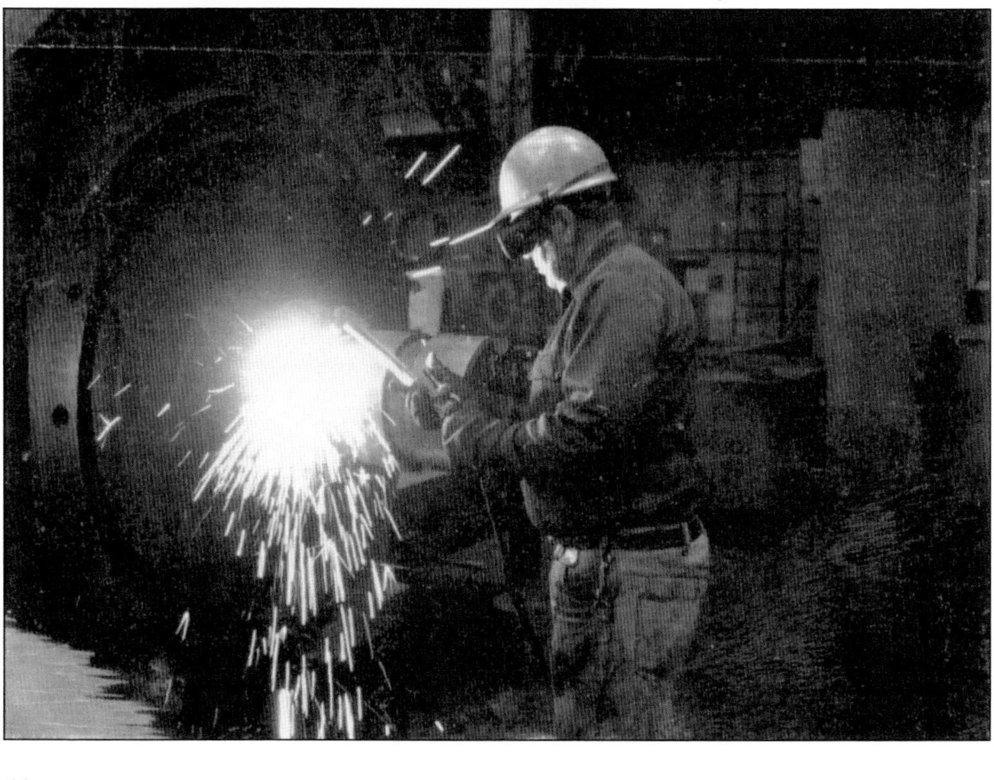

Dick Boess (standing), who worked for 28 years at Kaiser, was assistant supervisor of the Electric Weld No. 1 (EWI) pipe mill. He came from Evanston, Illinois. Jack Hoffman took his place when he retired.

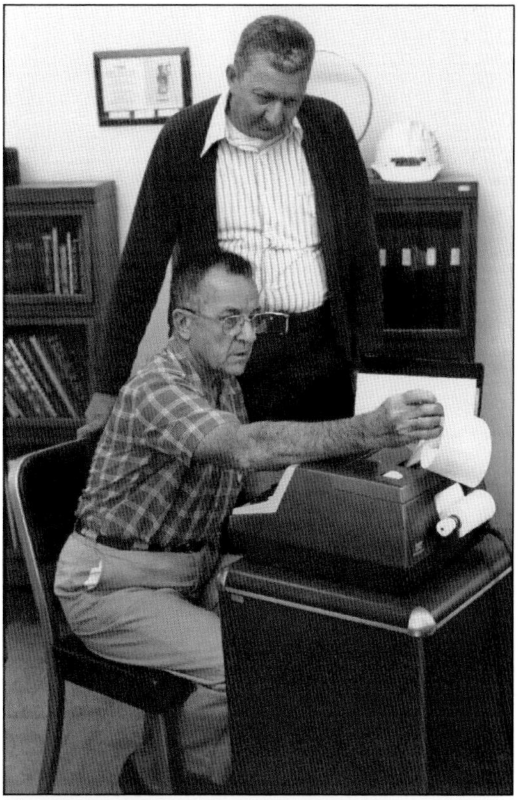

This photograph shows the EWI pipe mill with pipe coming out of the machine. The varied mills inside the overall "mill" included a structural mill that operated on the same principle as the plate mill but produced structural steel, and a merchant mill that also produced shapes and bars from carbon or high-alloy steels. Plates, structural steel, shapes, and bars could be made to meet almost any specification desired by industry.

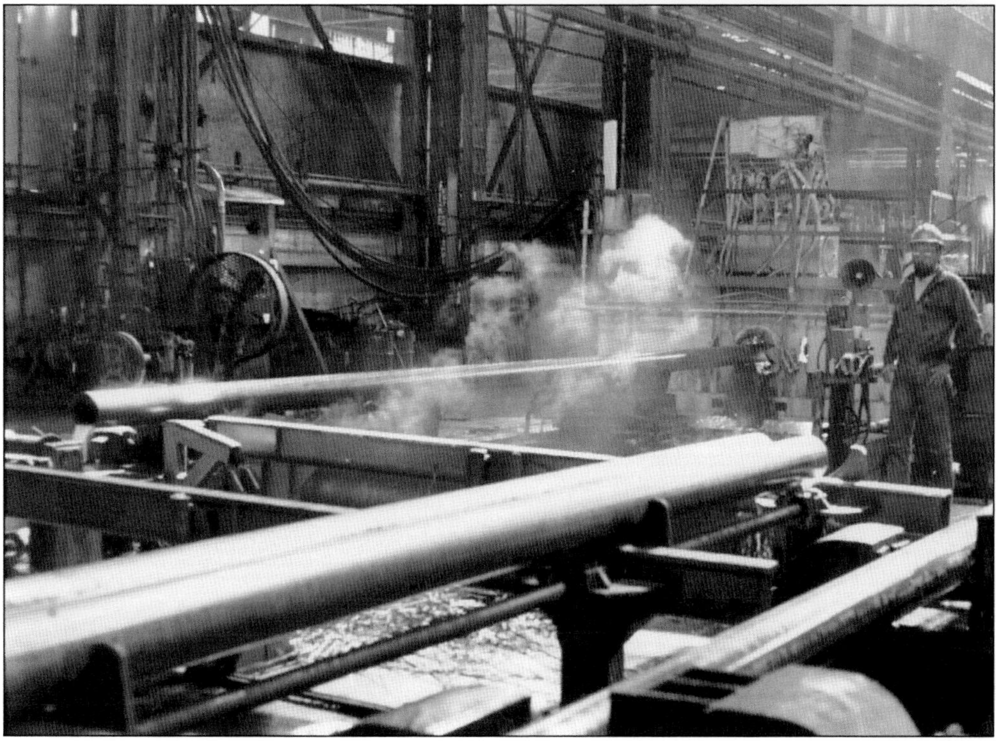

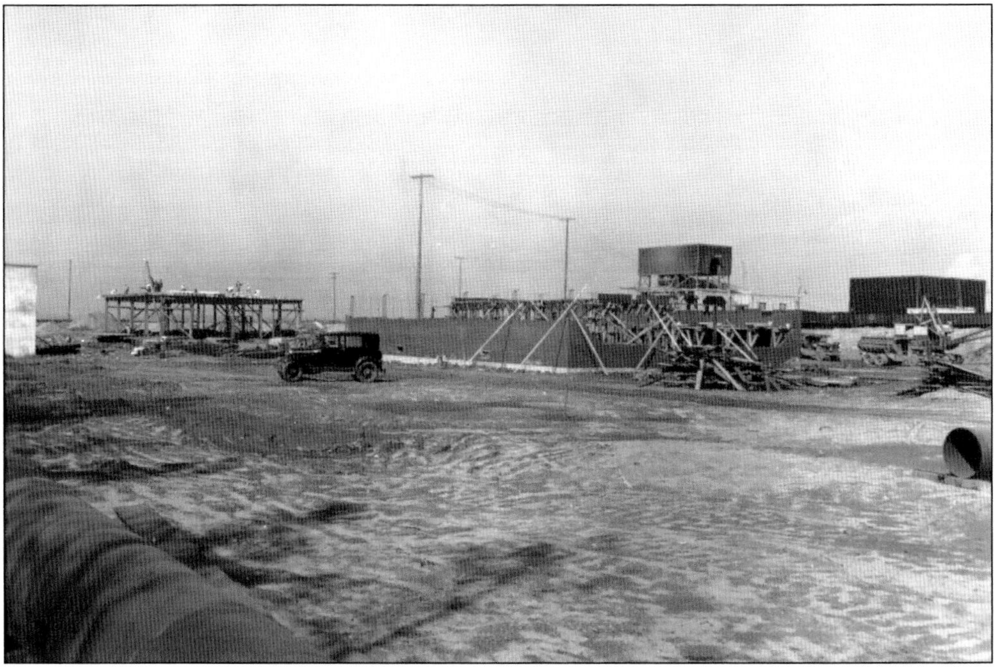

The pattern building and pattern storage building are under construction on February 27, 1943, in a view looking northwest. (Courtesy of Alexia King.)

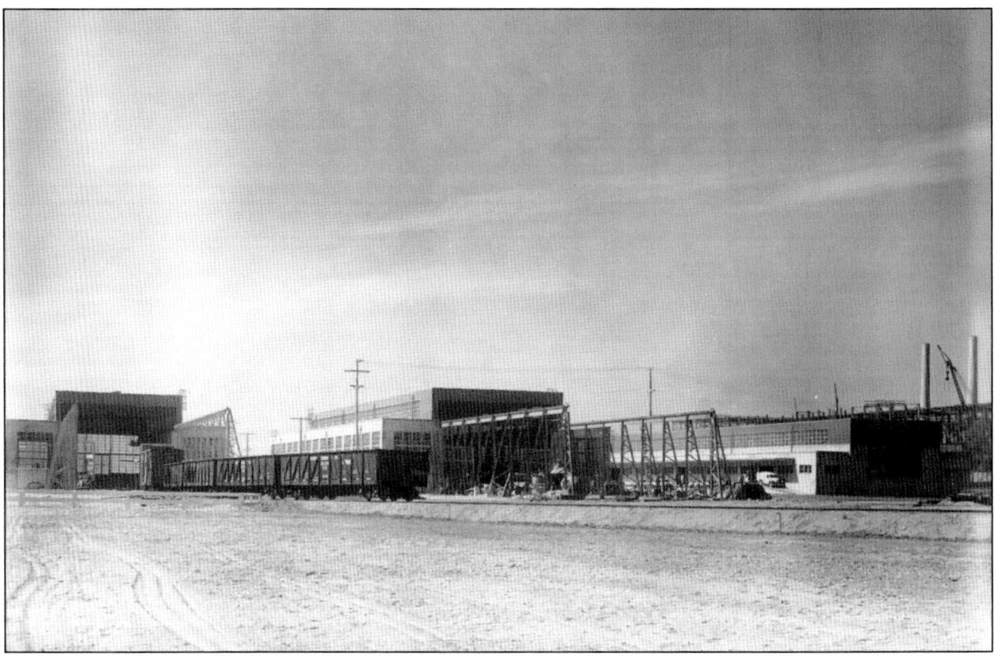

The services building area is pictured under some construction on February 27, 1943. (Courtesy of Alexia King.)

Mark T. Anthony, an executive at Kaiser Steel, worked at the mill for 34 years, which included two as president and one as vice chairman of the board. He hailed from Oakland, California. He is pictured here by a train of loaded steel cars leaving the yard.

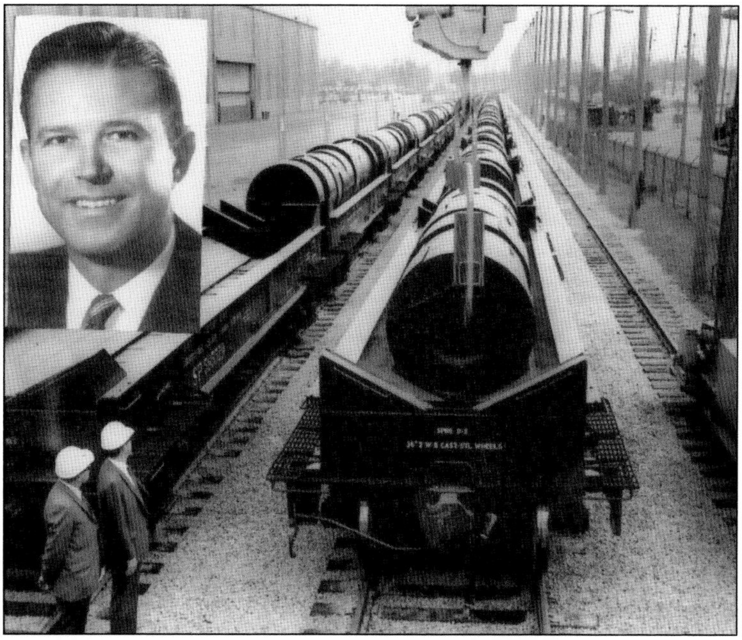

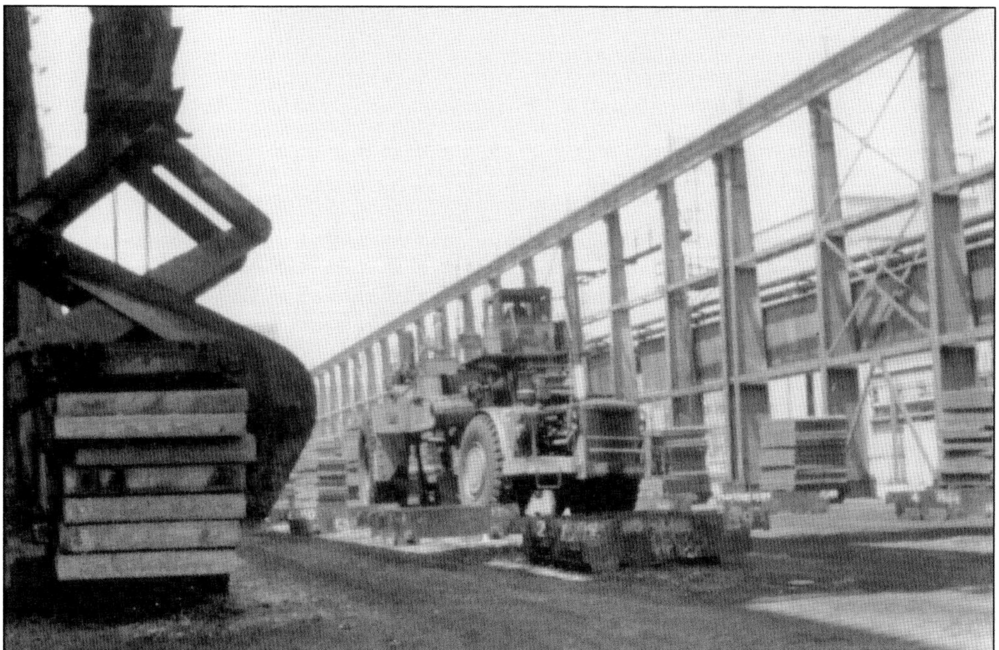

This is the slab hauler in the slab yard in 1977. A 100-ton machine, the hauler has since been cut up for scrap. Some of the workers that worked this yard are Daniel McCarthy, three years of service; Thomas McCarthy, 10-plus years in the EWI pipe mill and foreman of the tin mill; Billy Joe McKee, two years as a welder; Steven Lee McKee, two years as a crane-hook tender; Doyle Weaver, 20 years as an overhead crane operator; Norman McKee, 17 years as a riveter; Frank J. Burmingham, five years as a fitter layout man; and Jack White, 46 years in light steelmaking and treasurer for the United Steel Workers of America.

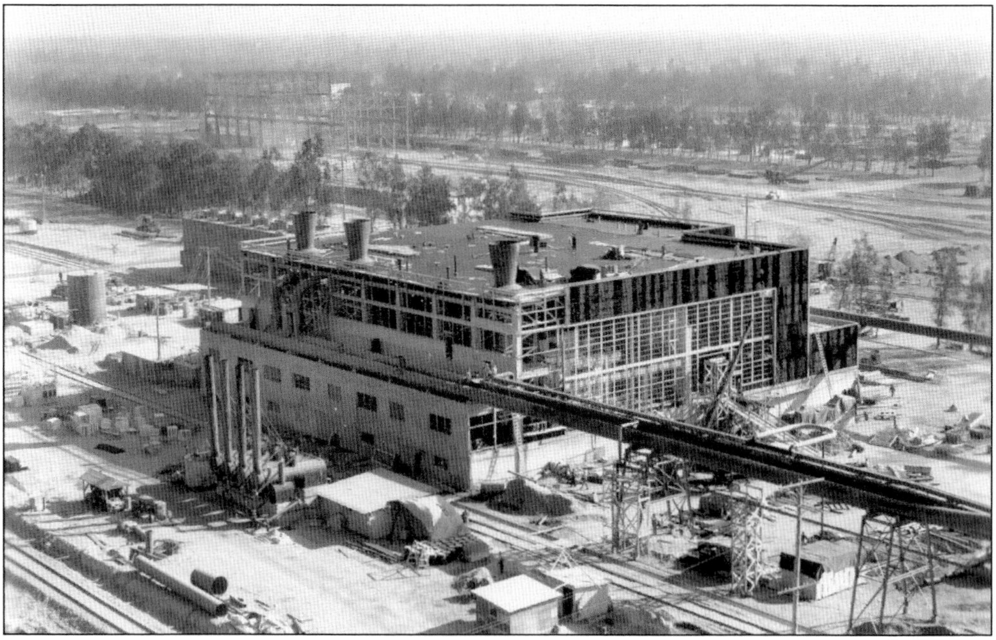

This is the powerhouse as Kaiser Steel was built in 1942.

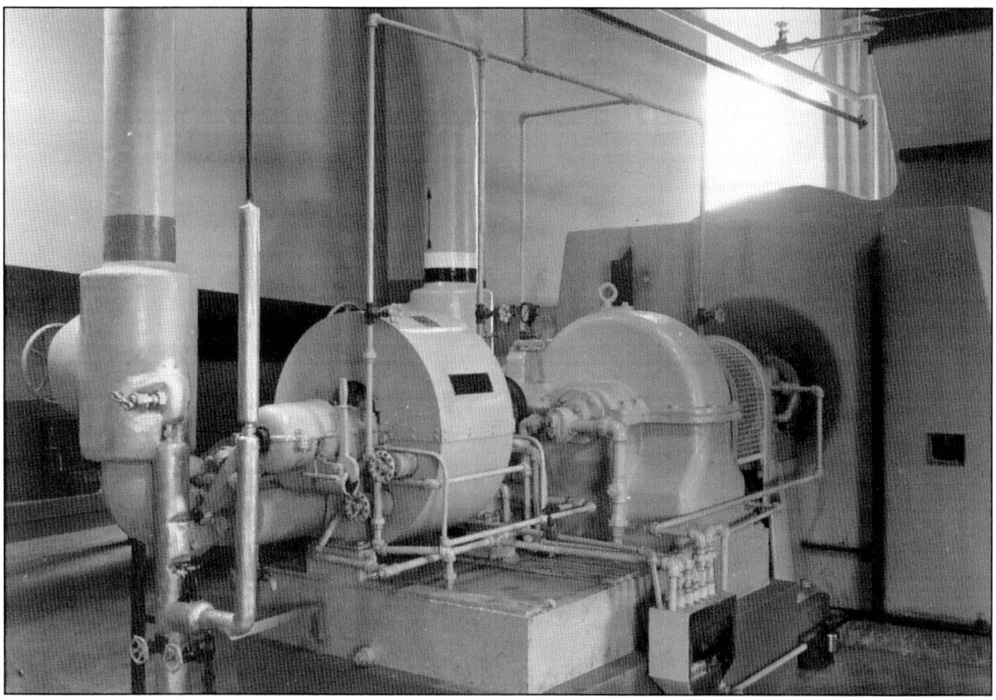

This photograph of the powerhouse's interior shows Boiler No. 3's inductive fan. The image was taken on February 27 1943. These workers in this area included Alex Templeton, plate-mill end-shear operator; Red Patterson, roller-plate mill worker; Red Fuller, plate-mill worker; Charlie Vall, plate-mill foreman; Arnold Ray McKee, 12-year machinist; Richard Alan McKee, 2-year painter; Elliott Morris, 4-year forge-and-welder worker; and Robert Schiemonz, 20-year Eagle Mountain and open-hearth worker who later welded on ships for Kaiser from 1943 to 1950.

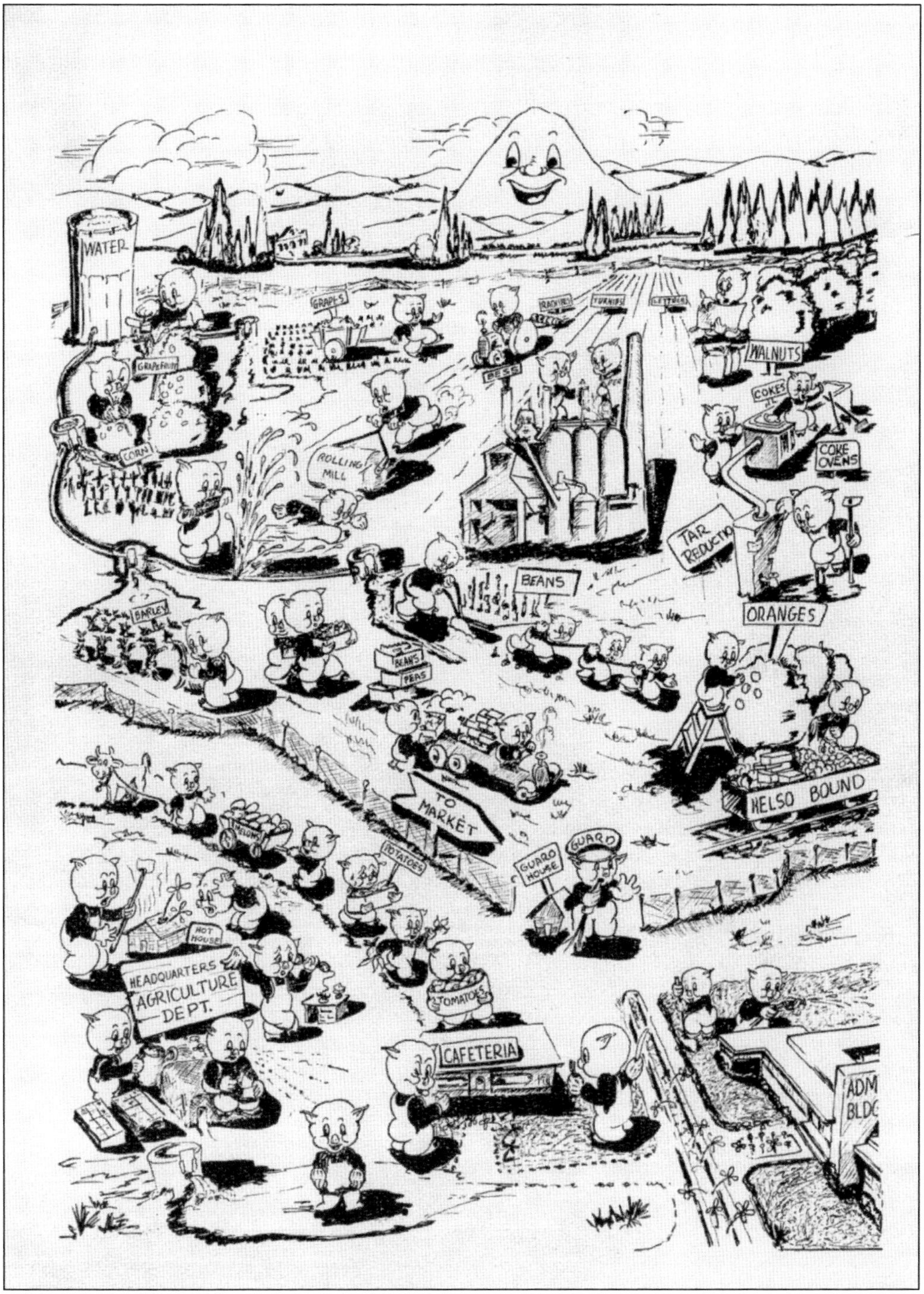

So much a symbol of Kaiser Steel, piglets present a tour of the agriculture-department area of Kaiser Company, Inc. An agricultural department would appear to have no connection with the making of steel and iron, but this one at Kaiser played an important part. With so large a plant covering more than 1,300 acres, so much of that land would be idle and unsightly if not for the productive crops or beautifying flowers, trees, and shrubs.

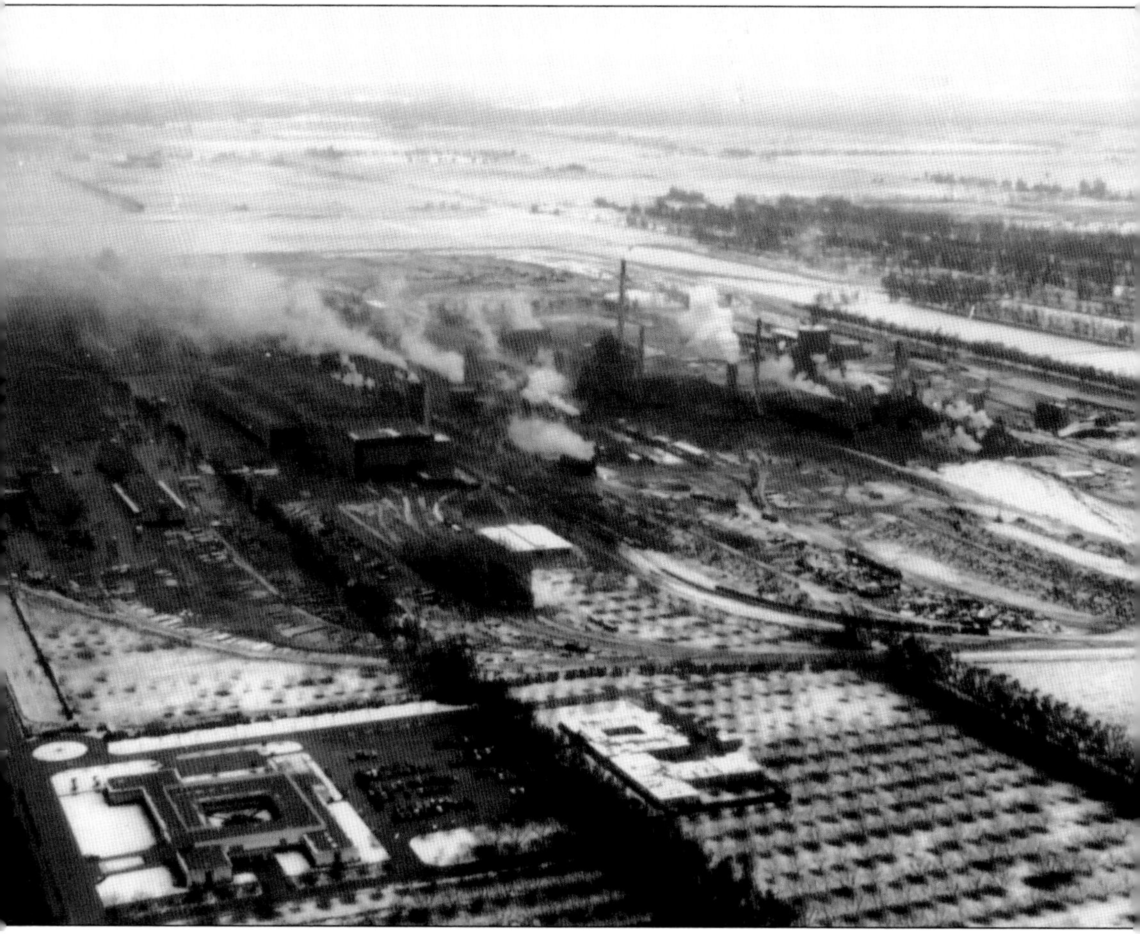

This aerial view of the Kaiser Company plant shows the surrounding community of farms and many different crops. When Henry J. Kaiser designed his mill, he not only planned an efficient plant, but one that would not produce the usual smoke and soot. He wanted his mill to be a giant of production while blending with the rural beauty of the community. During the war, the agricultural department was even more important for the production of fruits and vegetables, which were made available to the company cafeteria for employees' lunches and dinners or for sale on the open market. The last employee to work at Kaiser, Lee Armon (employed at Kaiser from 1964 to 1999), remained at the plant until 1999—17 years after it closed. He was in charge of the powerhouse and the water and sewer plant where the California Speedway is now, and he had to keep the waterworks going to provide water to build the grounds and speedway. The water tower, which was only used as a sign by Speedway, was gutted of equipment by Armon, who today lives in Corona. (Courtesy of Kaiser High School Library.)

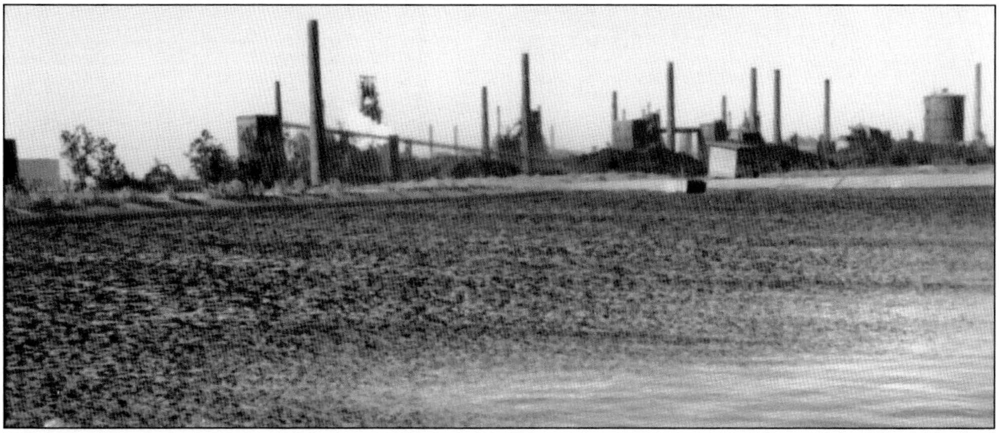

A freshwater pond, pictured here in 1971, was on the premises and inhabited by thousands of goldfish. Now gone, it was located on the northeast side of mill. With seven 730 acres of cultivated land encircling the steel mill, virtually every crop native to the locality was grown in the postwar era, including 96 acres of vineyards, 150 acres of walnut trees, 400 of hay and grain, 23 of yams, 12 of black-eyed peas, five of corn, 15 of citrus, 5 of tomatoes, 2 of potatoes, and 18 acres of other garden products. All the permanent buildings were landscaped with velvet-smooth green lawns, evergreen shrubs, and a constant succession of blooming flowers. A blaze of color was maintained in the flower borders by using plants raised from seed in the Kaiser Company hot beds and lath house. The attractiveness of the plant was known regionally.

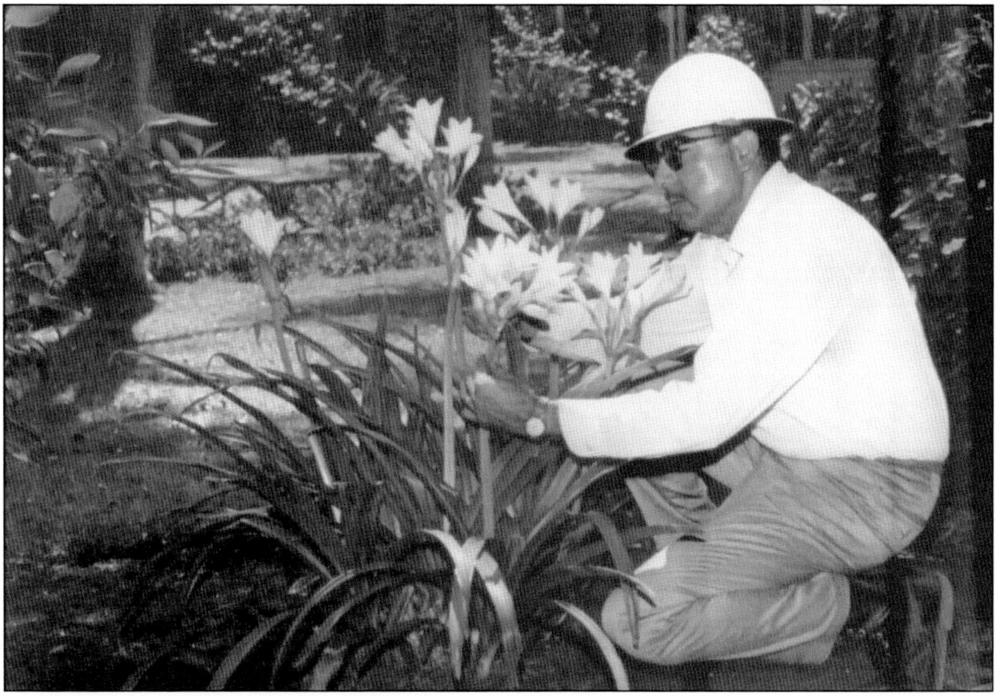

Frank Moran, a gardener on the agriculture crew, carefully examines a beautiful amaryllis plant growing in the administration-building patio to make sure it is receiving proper nourishment and cultivation. Kaiser Steel has been called the most beautifully landscaped steel mill in the world. Gardeners such as Frank were responsible for landscaping as well as weed and pest control. This photograph was taken by Timon "Curly" Covert for a 1976 calendar.

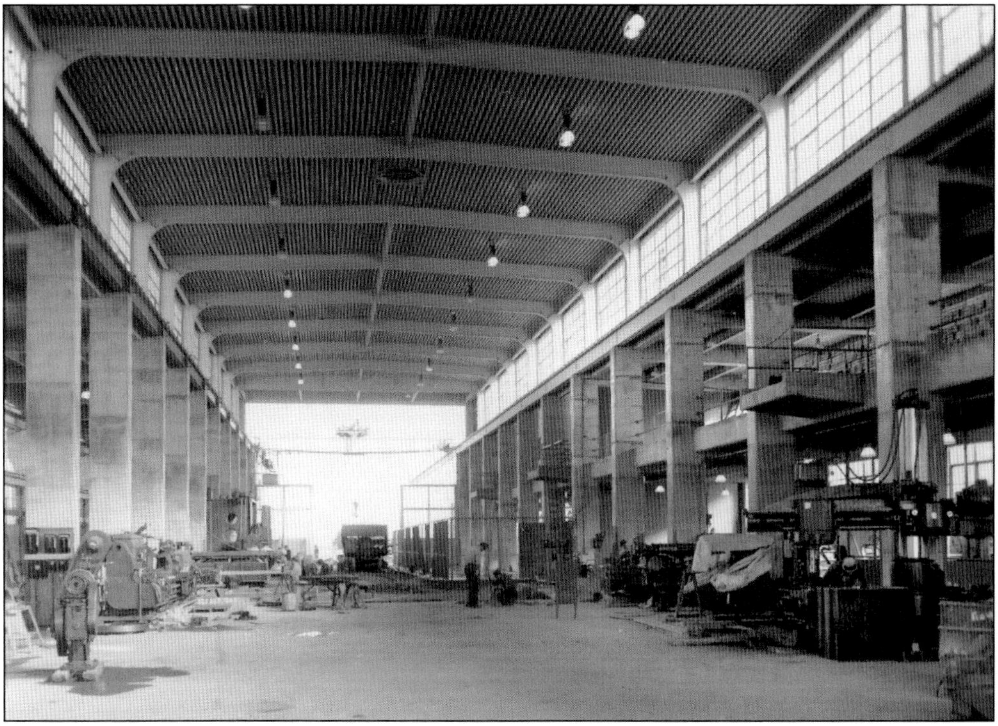

This interior image of the machine shop building looks east and was taken on February 2, 1943.

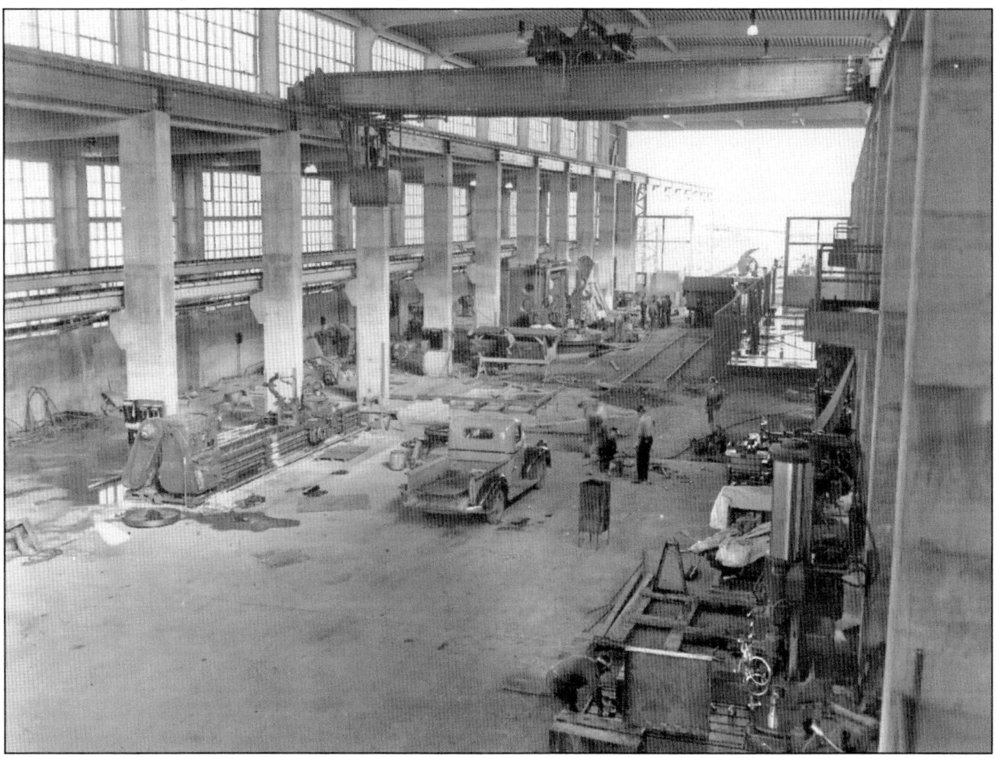

This is another view of the machine shop looking east. (Courtesy of Alexia King.)

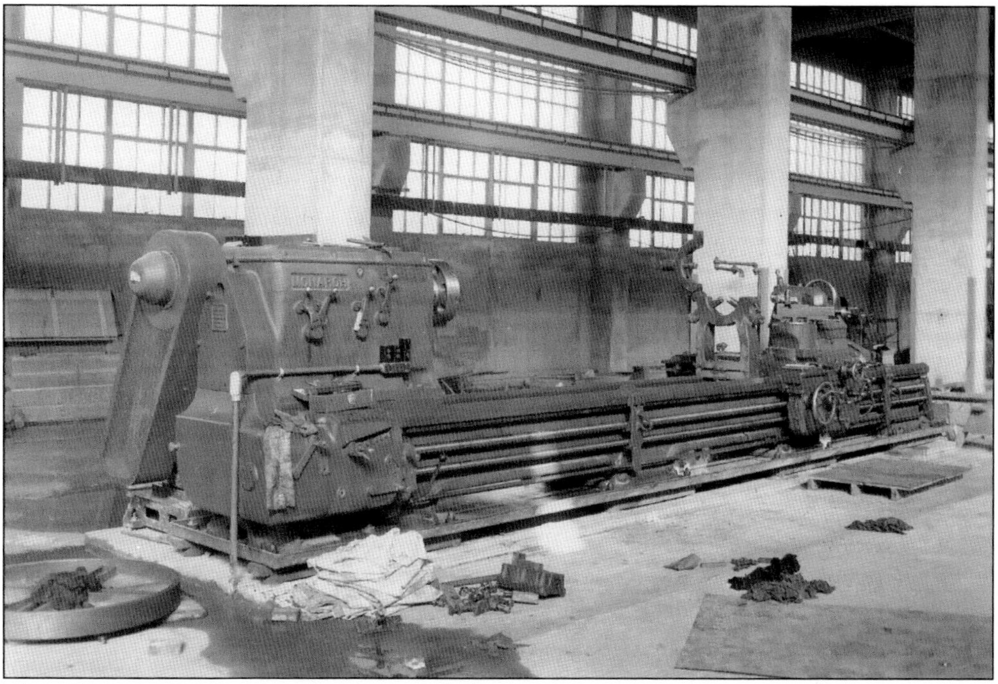

This photograph of a 22-inch Monarch lathe was taken on February 2, 1943. (Courtesy of Alexia King.)

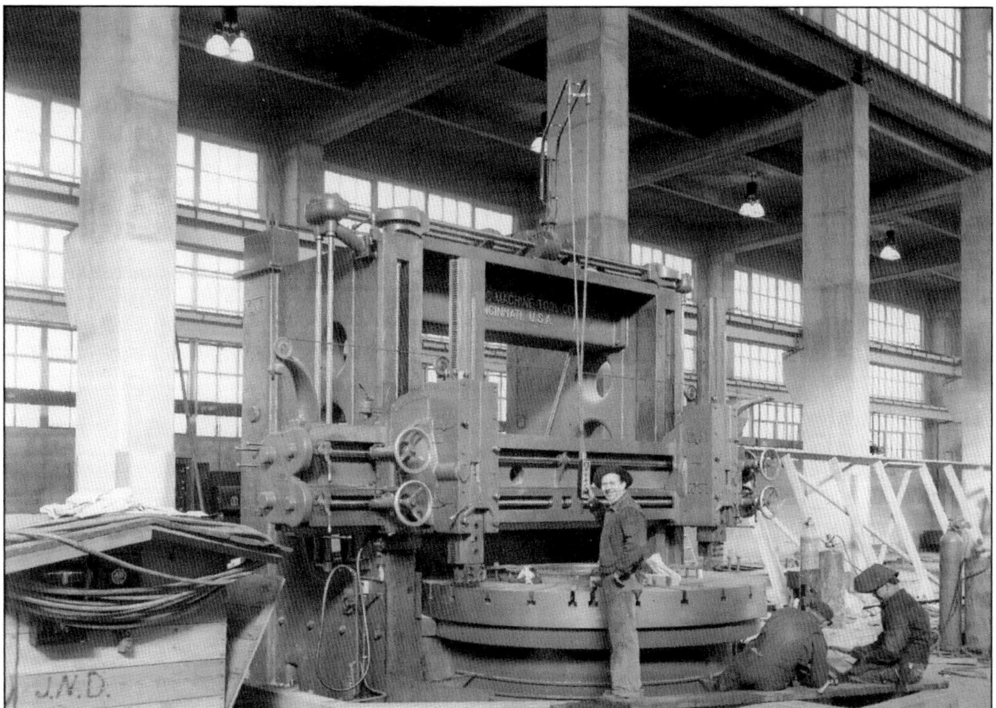

This structure was known as the Cincinnati vertical boring building. The photograph was taken around February 3, 1943. (Courtesy of Alexia King.)

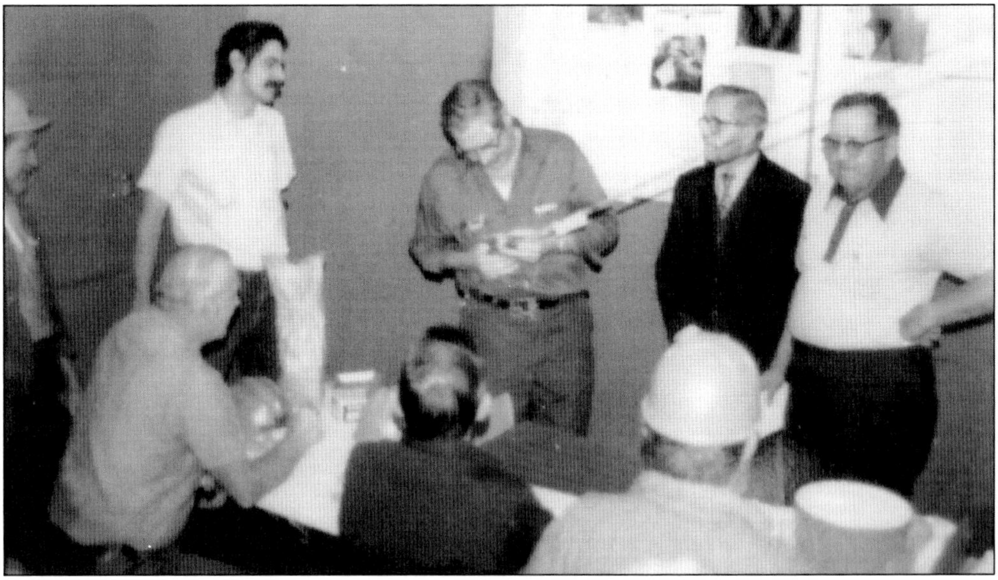

Lloyd Husted, who worked for 32 years in the machine shop, takes it all in on his retirement day. Pictured here, from left to right, are Lloyd's son, Larry Husted, Mayor Trudeau Lallenson, the back of Little, and the back of Cliff Morgan. Lloyd is pictured center.

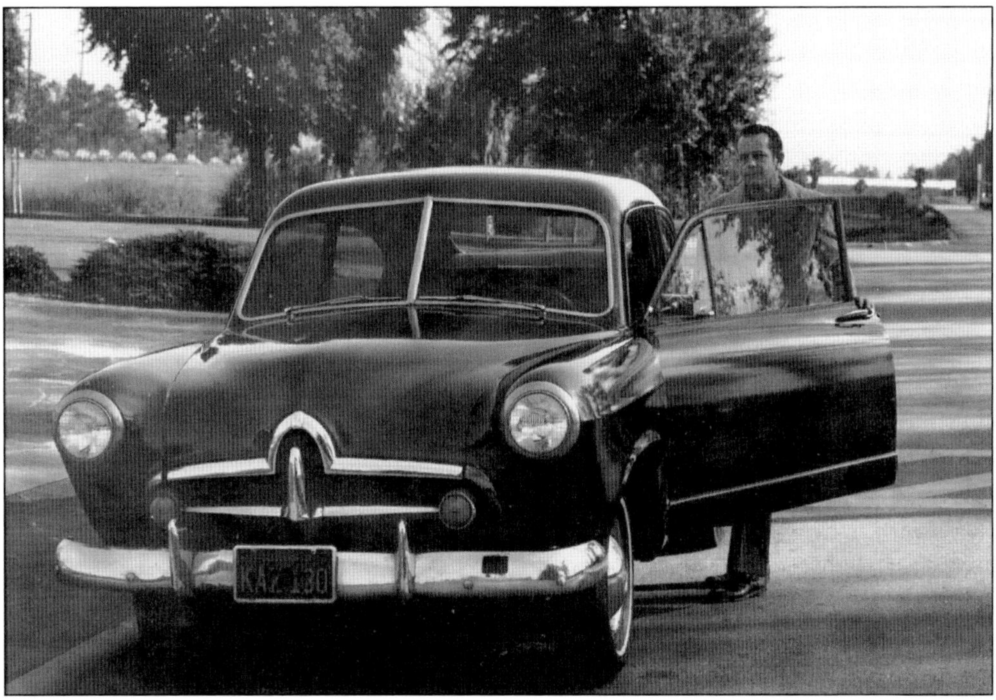

Howard Herman "Twigs" Jordan worked for 32 years as the supervisor of the west-end maintenance department. He also worked for 10 years at California Steel. He owned for many years an original Kaiser car, although with more power than it was originally built with. He showed it at the plant when Henry J. Kaiser wanted to see it. Jordan hailed from St. Louis, Missouri. He donated a pig-iron and sample raw-materials display to the new Kaiser Museum being organized in the city of Fontana.

Plant cranes must meet certain safety standards, and it is part of the crane repair crew's responsibility to inspect and service each one. Checking this ingot-mold crane are Bob Collard, Ed Wellington, John Sturgess, and supervisor Marv Freeman. One crane operator, Joseph Trebizo, worked for 33 years.

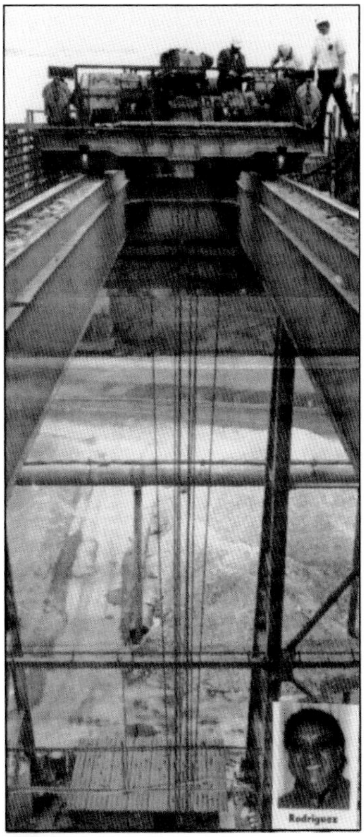

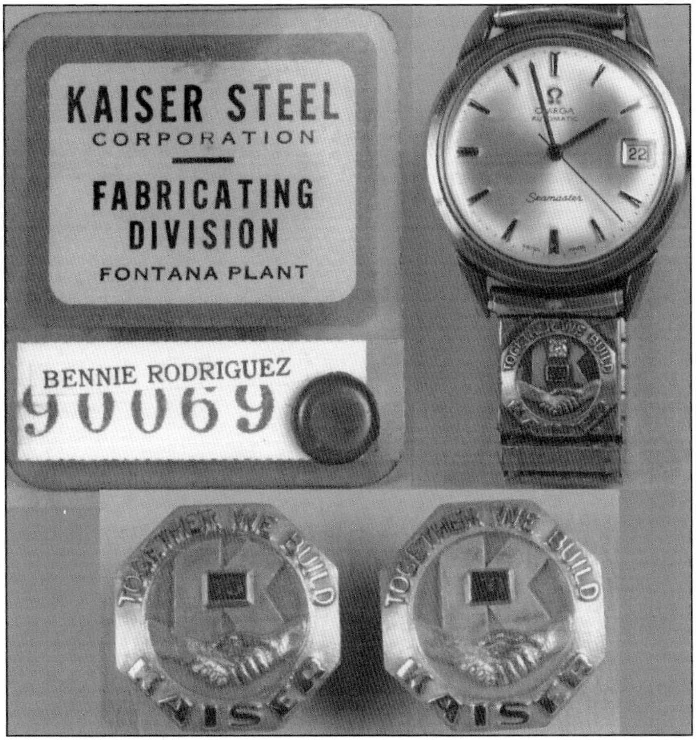

Benjamin Rodriquez (inset of above image), who hailed from Texas, worked for 31 years as a crane operator for Kaiser Steel. Some of the gifts that Kaiser bestowed on him are pictured here with his badge. (Courtesy of Jim Rodriquez.)

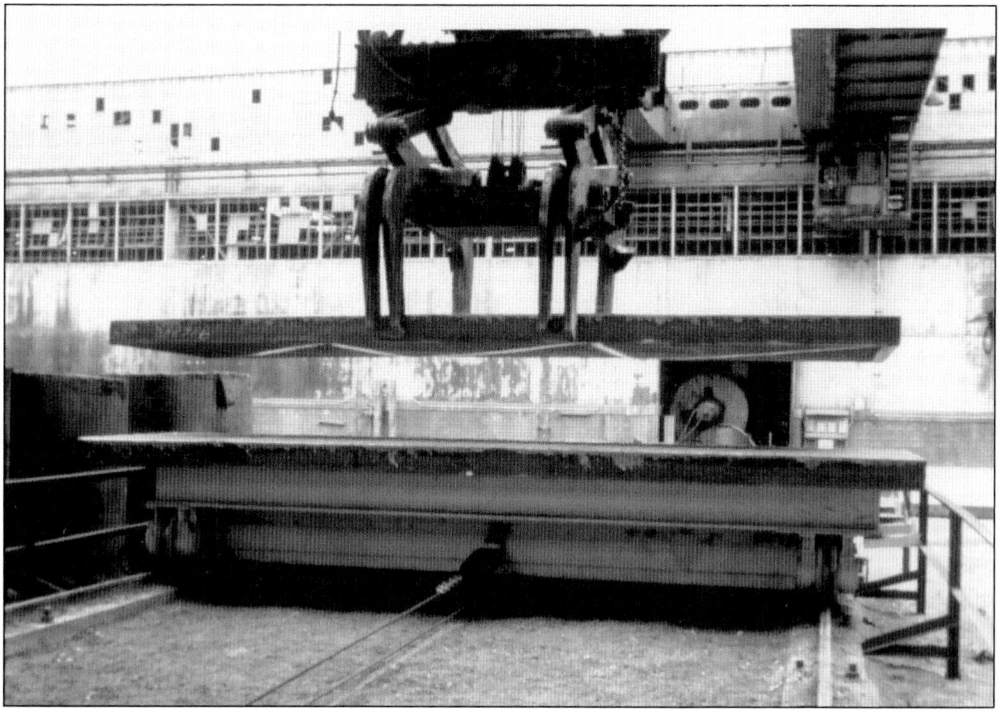

This photograph demonstrates the size of some of the items lifted by Kaiser Steel cranes.

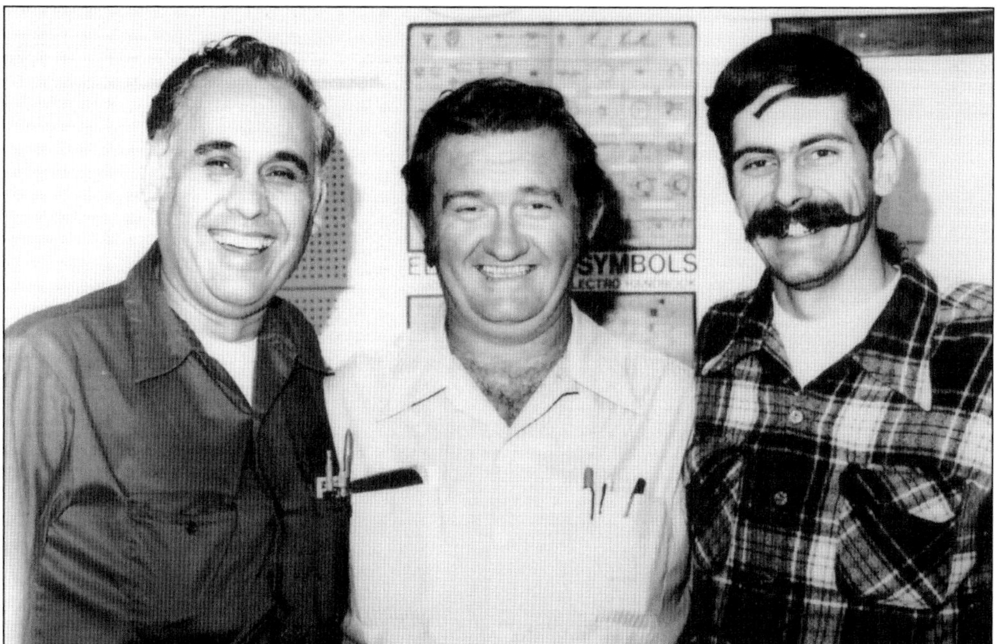

This is an image of the crew that worked in the instrument shop. These workers, from left to right, are Felix Peraza and Mike Popick, both instrument men, and Larry Husted, an instrument apprentice who worked in the department for 15 years. He came from Cedar Rapids, Iowa, as did his father. (Courtesy of Larry Husted.)

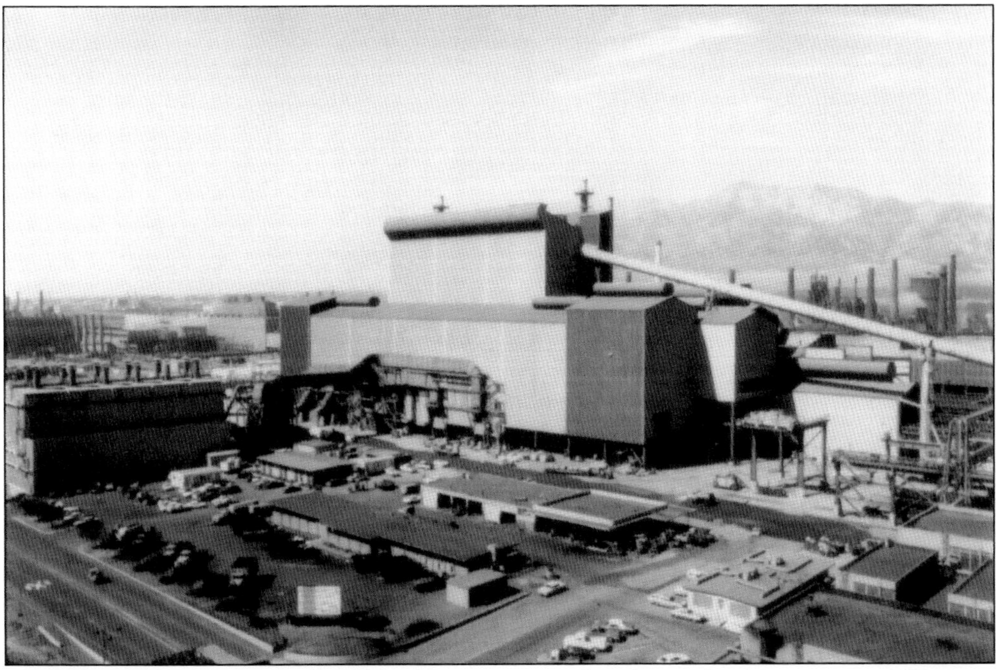

The basic oxygen-processing (BOP) building was central to the steelmaking process at Fontana. It is pictured here at its dedication.

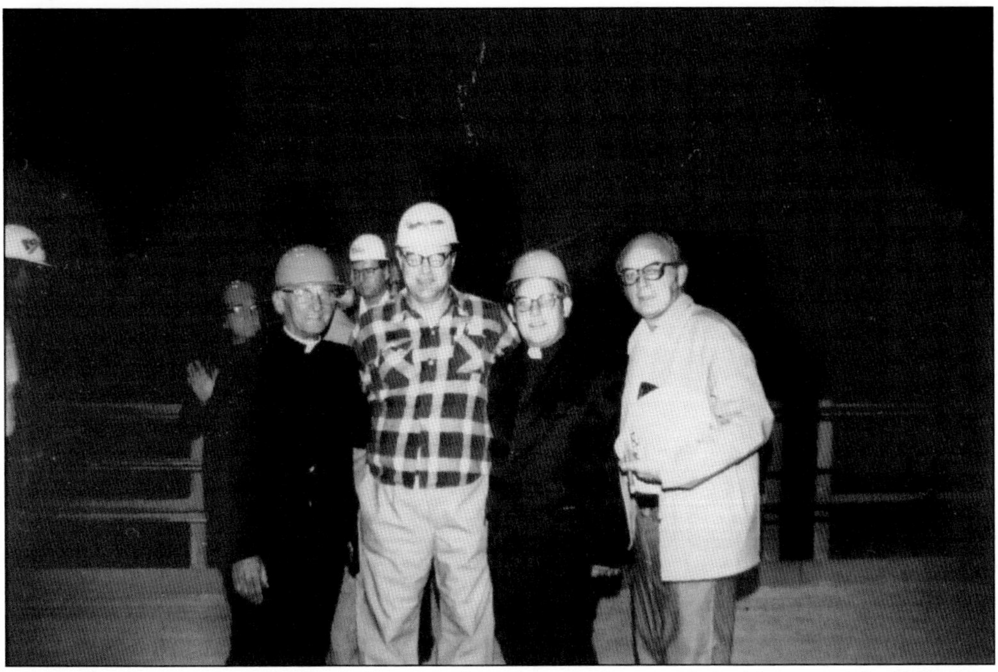

The priests from St. Joseph's Catholic Church in Fontana bestowed the blessing of the BOP. The Reverend Father Vincent Grzybowski and an assistant handled the honors. The gentleman at in the photograph is John J. McCarthy, who worked for 24 years in various Kaiser executive jobs. Originally from Chicago, he had a reputation for being especially adept at troubleshooting. The gentleman in the center is James Szilagyi.

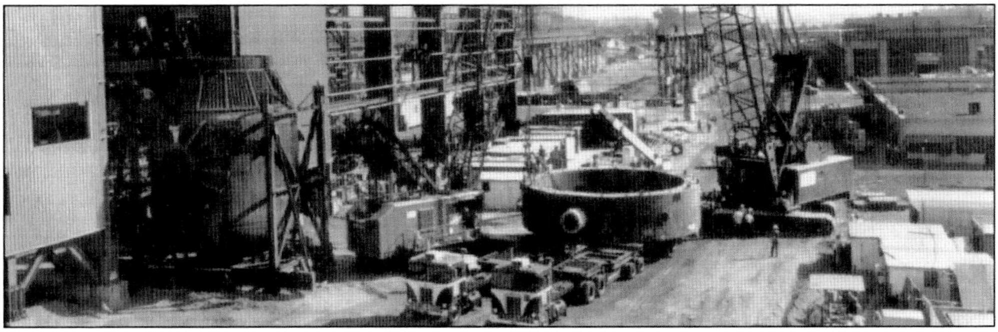

Howard Craig became the supervisor and assistant manager of transportation for Kaiser Steel. He joined Kaiser in 1951 as a rate clerk in the traffic department. He became a traffic analyst in 1956 and that same year was promoted to supervisor of traffic, a position he held until his appointment as assistant traffic manager. He was then made responsible for providing traffic and/or transportation services for all outbound shipments from Fontana. His biggest project was when he went to see NASA's first space shuttle when it was rolled out of a California hangar at Edwards Air Force Base, for he had been stumped as to moving a piece of equipment for the BOP building. When he saw how the shuttle was supported on wheels, it inspired the following solution pictured here.

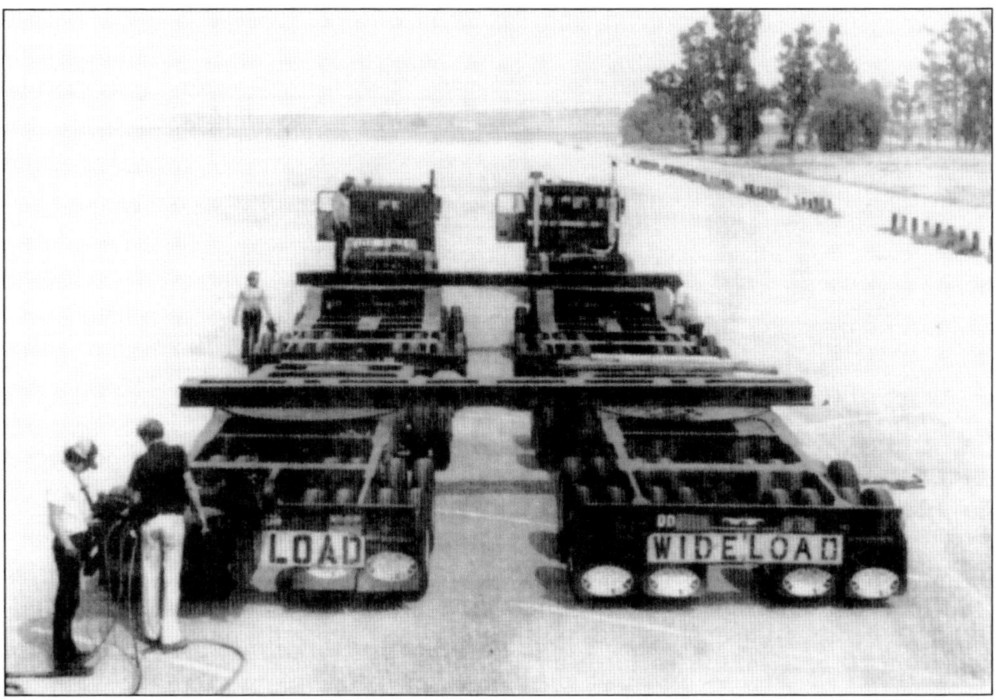

Contractors Cargo Company built two special trailers to transport the two 217-ton BOP furnace trunnion rings to the Fontana mill. The trailers are pictured being connected with steel beams in the parking lot of the Ontario Motor Speedway. Pylons were set up so that drivers of the tractors could practice maneuvering together under conditions they would encounter in hauling the rings.

Personnel of Contractors Cargo Company performed precise measurements in loading a 217-ton basic oxygen-process furnace trunnion ring for shipment from the plate fabricating plant in Fontana, on Slover Avenue, to the steel mill. The two 225-ton-capacity furnace vessels and structures for Kaiser Steel's $233 million modernization program were fabricated and erected by KSC's metal products and fabricating division.

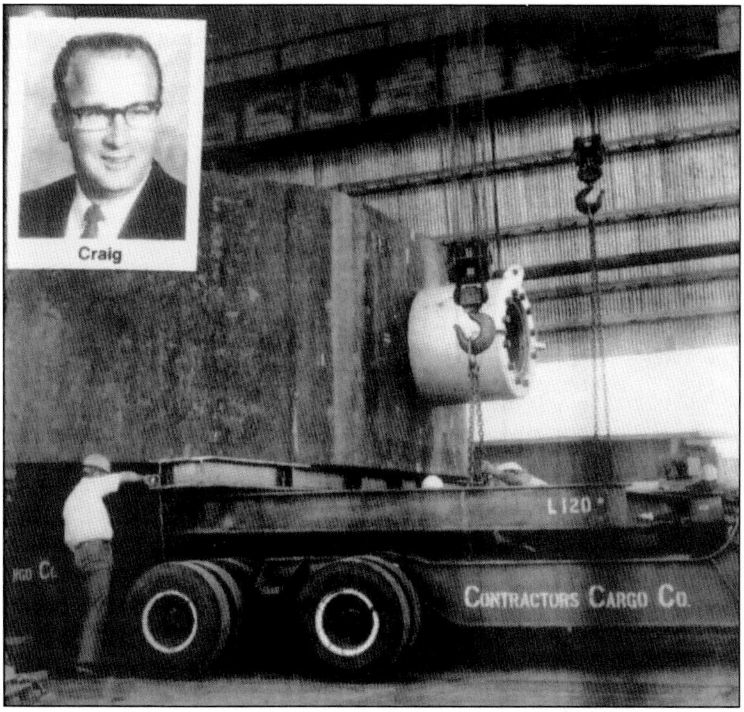

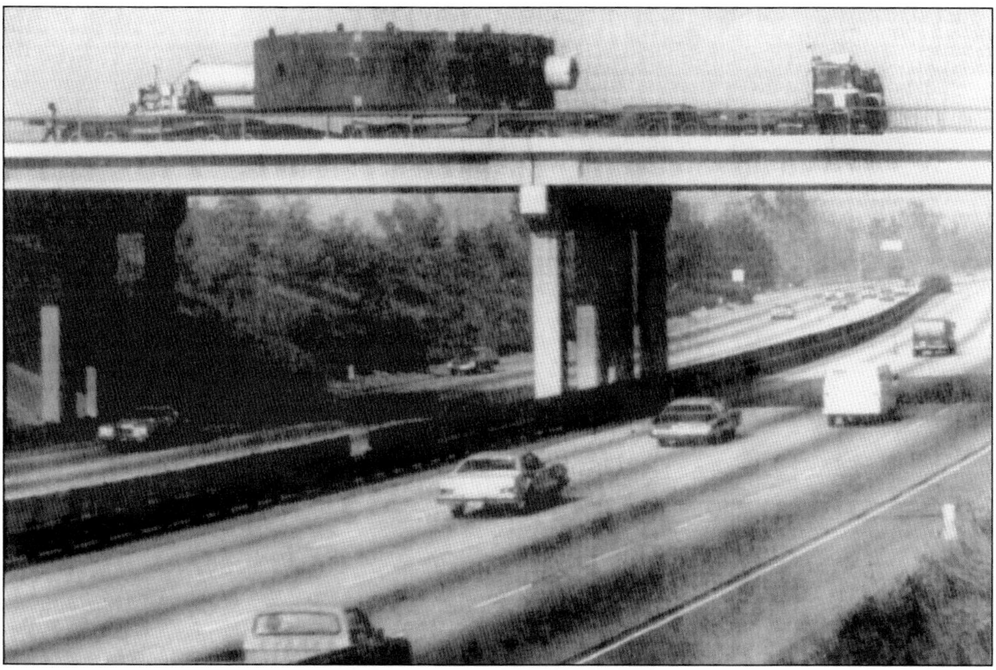

On August 25, 1977, the 217-ton BOP furnace trunnion ring, pictured here, was transported over the Cherry Avenue overpass on Interstate 10. With the tractors and dollies, the entire load totaled 277 tons (554,000 pounds on 148 wheels), making it the heaviest load ever to cross a California bridge.

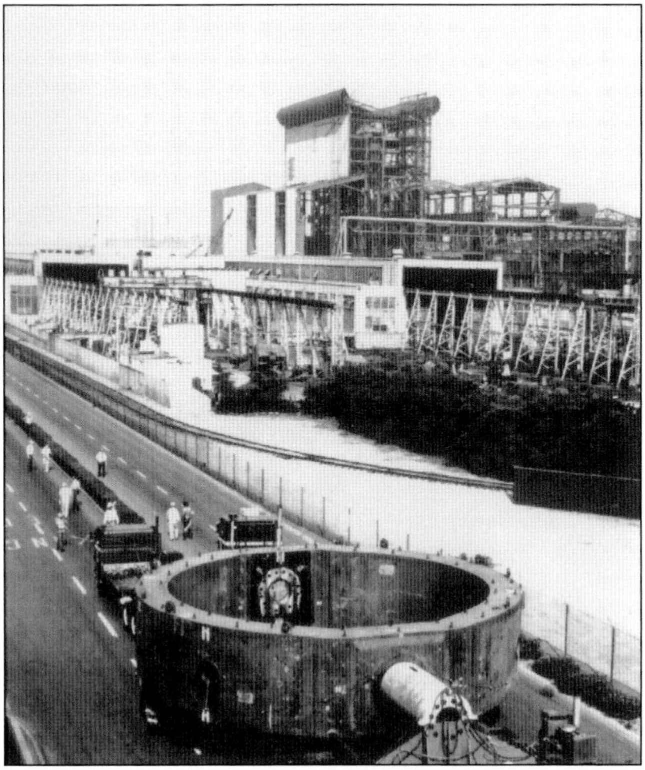

Bob Howard told the International Trade Commission that Kaiser Steel was doing what it could to meet competition from imported steel in the West Coast market. When the modernization was completed, the company had a competitive modern steel plant. The second BOP furnace vessel awaited its trunnion ring here, which is pictured aboard two tractor-trailer rigs transporting it to the job site.

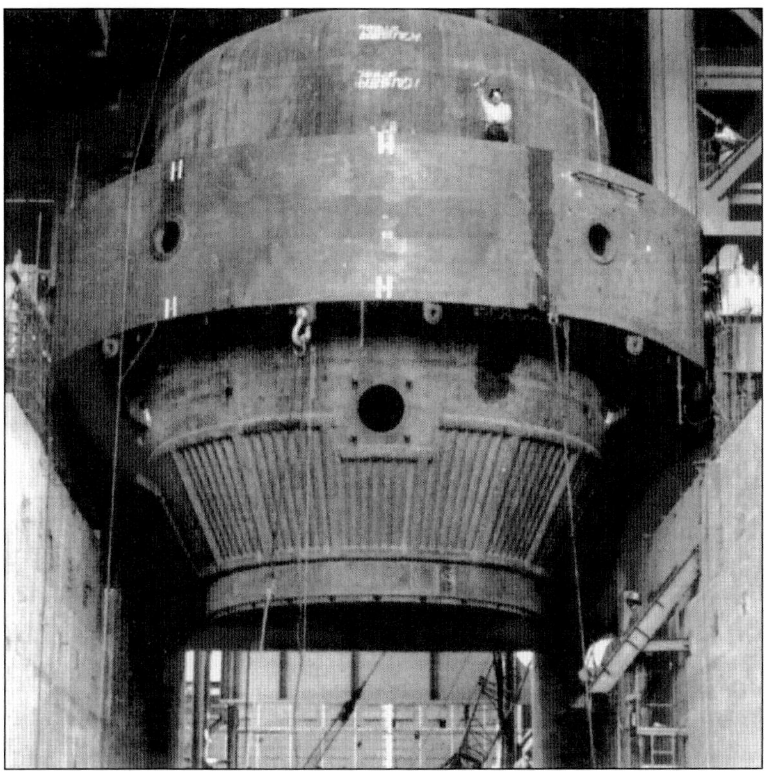

The No. 5 furnace for the new basic oxygen-process steelmaking shop at the Fontana mill is pictured here upside down. This allowed construction crews to fit the keys into the keyways between the main gear box's bull gear and the trunnion ring shaft. The electrically operated gearbox was then able to rotate the furnace during the production process.

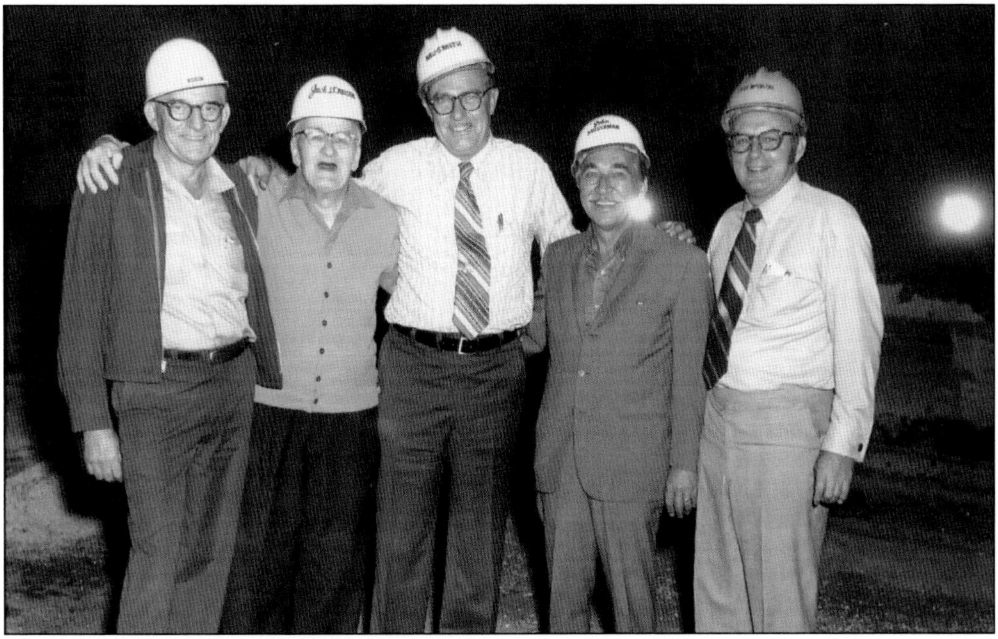

These people involved with the basic oxygen process in the mill, from left to right, are James Szilagyi, Jack Issac, M. J. Smith, John Saussaman, and John J. McCarthy.

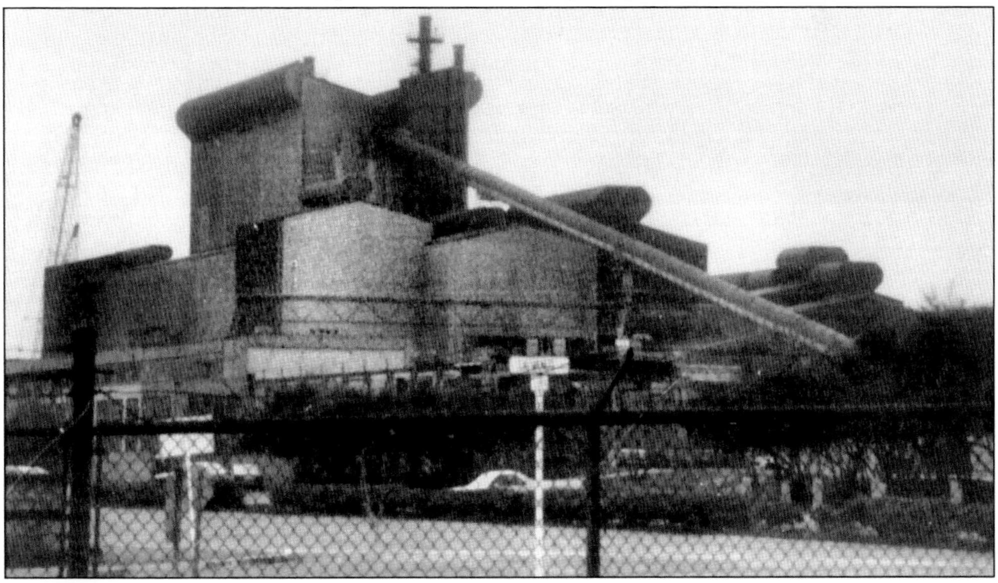

The BOP building is pictured here at the beginning of its dismantling by Chinese engineers, who supposedly reconstructed it in China. The casting building was also taken down. Three hundred Chinese arrived to do the work, numbering each piece as the building was taken apart. California Steel Industries, which bought the plant for $100 million from Kaiser Steel in 1984, sold the mill to the China Metallurgical Import and Export Corporation of Beijing for $15 million. It took two years to take it apart. The No. 2 basic oxygen-process and caster facility opened in 1978 at a cost of more than $250 million and was capable of melting and shaping 700,000 tons of steel annually. It operated for five years before the bottom fell out of the domestic steel industry. Increasingly restrictive air-quality laws delivered the knockout punch. (Courtesy of Bert Bene.)

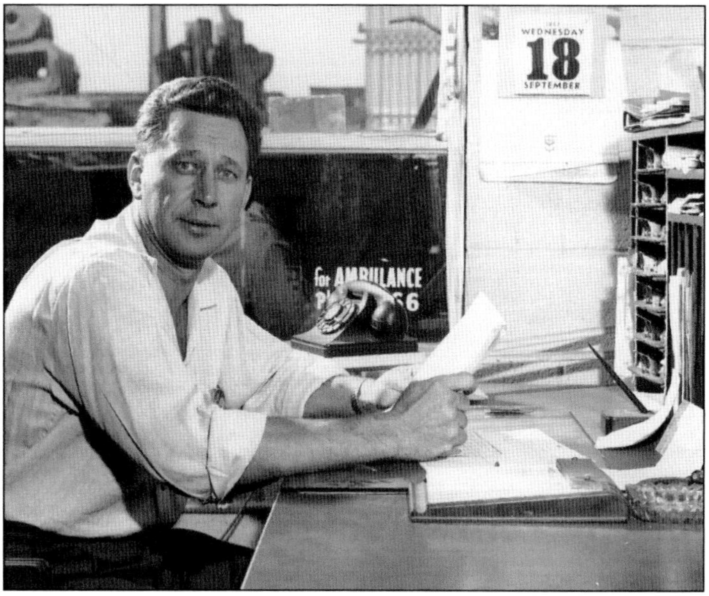

Frank Horzen worked for Kaiser Steel when his family started to grow. He had been a baseball player, but changed jobs for the sake of his family. He spent many years in the paint and carpentry shops and later became interested in Fontana politics. He was on the city redevelopment staff and later served two terms for a total of eight years as mayor of Fontana. His hat carried his nickname, "Lefty."

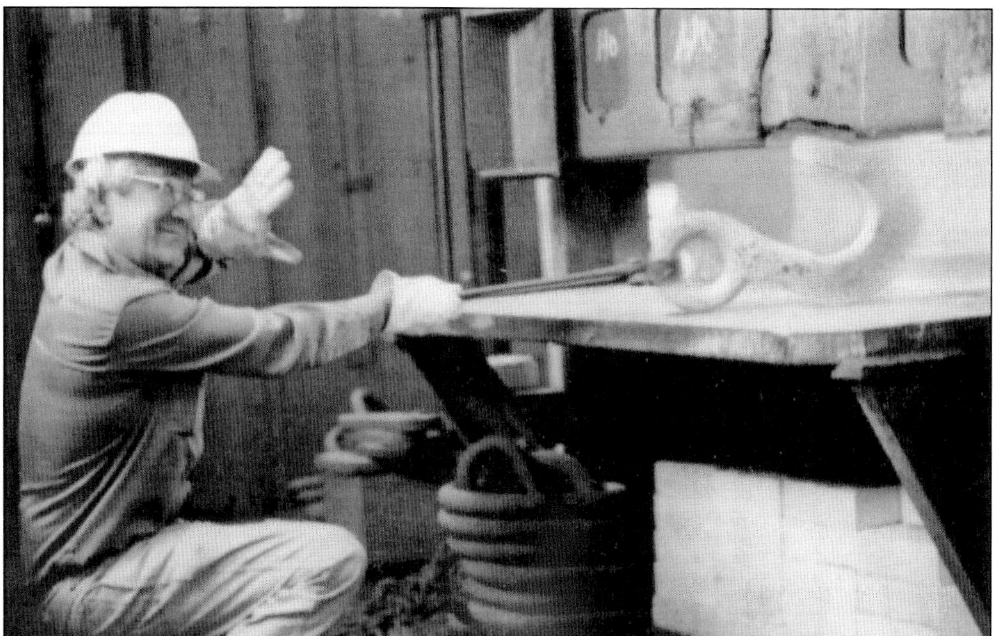

Timon "Curly" Covert was another person who came to California from Colorado. He started working in 1951 as a laborer in the EWI pipe mill, then later to form and weld in the same department. He was, in 1953, elected to the zone 14 grievance committee. He was appointed chairman of the plant-wide incentive committee in 1957, and five times served as a member of the contract negotiating committee, including the 1973 contract. He became active in local civic affairs, serving as Fontana city councilman for two years, and on the City Planning Commission for seven years, with the final year as chairman. He became interested in photography and took a lot of images of Kaiser for two calendars. Many found their way into magazines.

The Flores family consisted of, from left to right, Ben, who worked 25 years in the plate mill; Juan, who toiled 30 years in the specialties department; Ernie or "Danny," who was employed for 10 years in the plate mill; and Manuel Honesto and Richard Valadez II, both of whom worked for 30 years at the mill (not pictured).

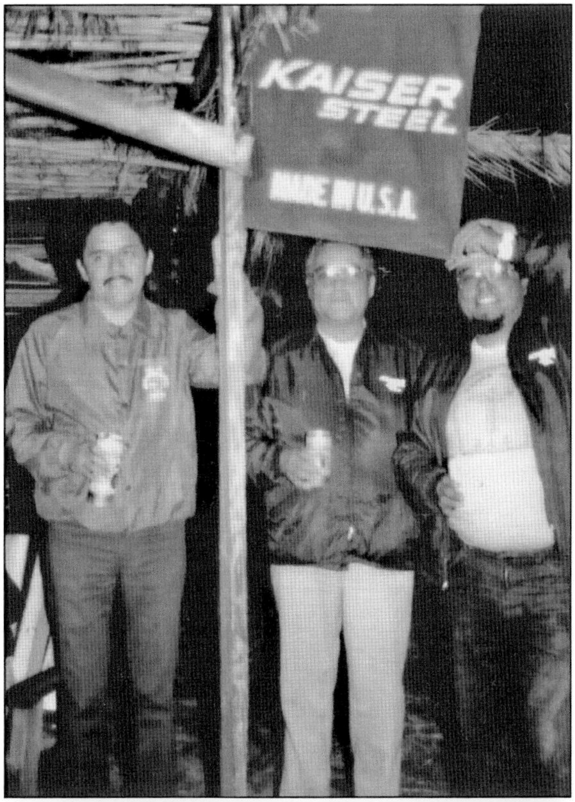

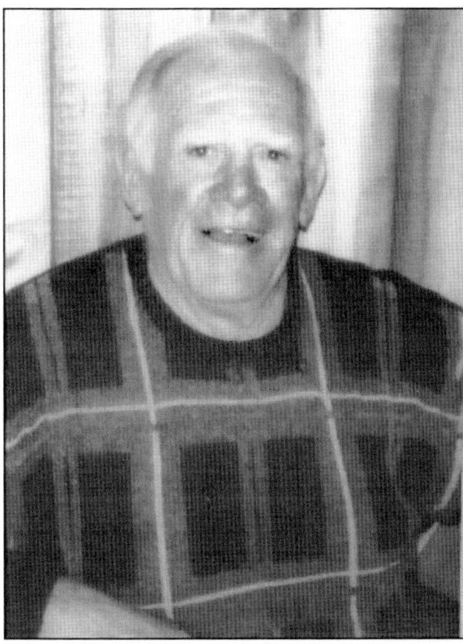

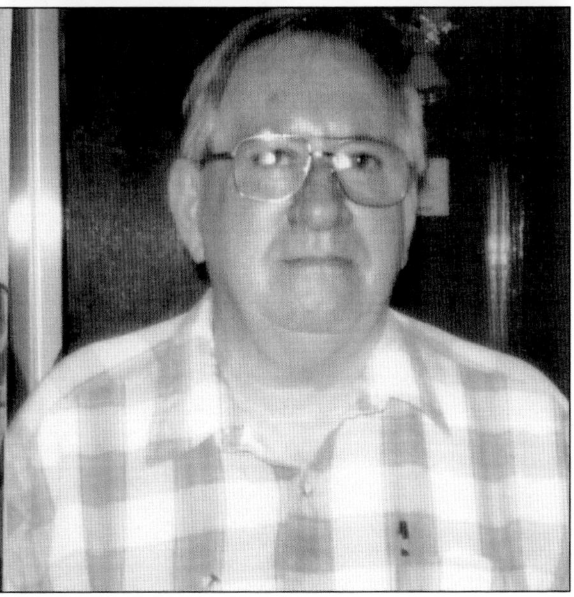

The following two persons are on the New Fontana Heritage Museum Association, working for a Kaiser Museum. They are old-timers who worked for Kaiser and later for California Steel. Edward Cook (left) spent 29 years at Kaiser and 11 years for California Steel as supervisor of material handling for both, and Charles Dall was a railroad switchman for Kaiser for 14 years before going to California Steel where se is currently a hauler.

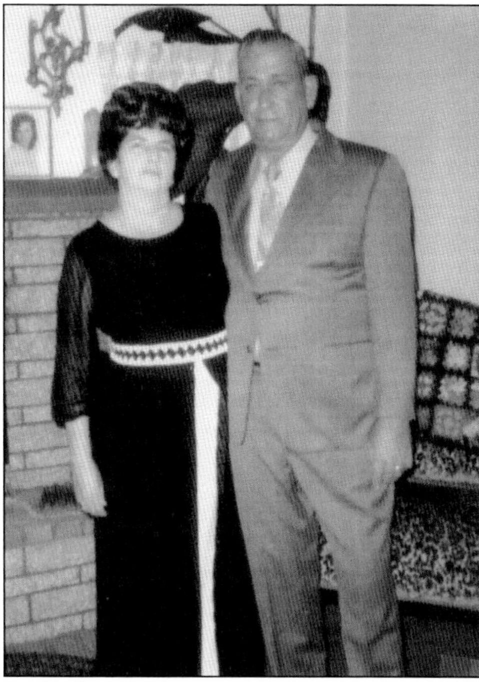

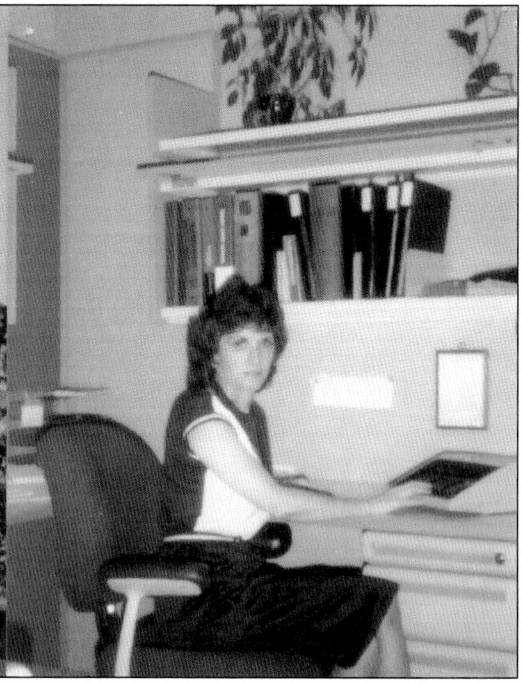

The Long family all worked at Kaiser. John Long worked for 33 years in the hot-strip mill as inspector. His wife, Ruth, was a tin-mill inspector for 20 years. Their daughter, Jonnie Long, worked for 25 years as claims adjuster, with two of those after Kaiser closed its doors while she cleared out the ledgers. She later went to work for California Steel for eight years doing the same thing. The grandchildren also worked there while attending college. The Long family originally came from Maryville, Tennessee.

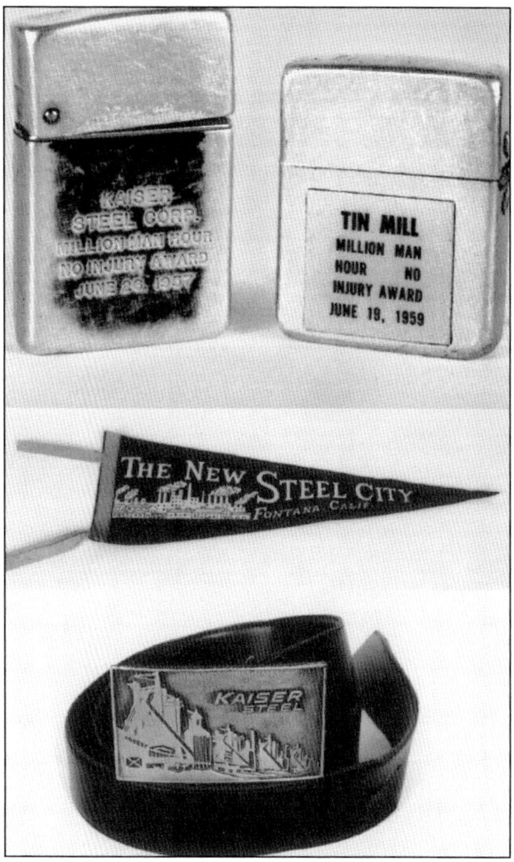

The items in this photograph are some of the incentive gifts given to employees when no accidents occurred per 1,000 man-hours of work. (Courtesy of the Long family.)

Four

Kaiser Organizations

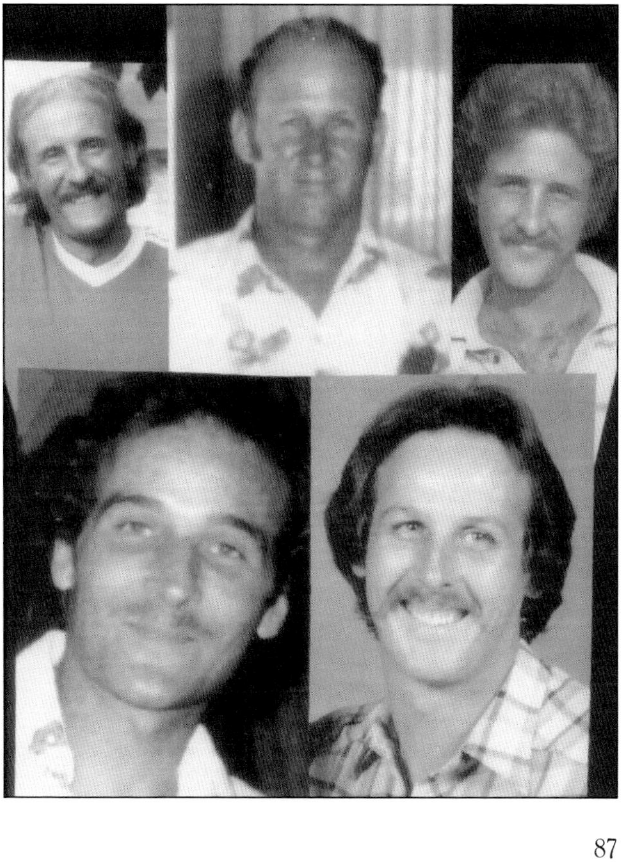

Billy Gene McKee (top center) was a worker in the Kaiser fabrication division for 22 years. He was also the president of the United Steel Workers of America Local No. 7451. He represented the chapter at many out-of-state meetings. He was killed in a jeep accident on a hunting trip. He had four sons who also worked at Kaiser. Clockwise, from above left, are Billy Joe McKee, Billy Gene McKee, Richard Alan McKee, Arnold Ray McKee, and Steven Lee McKee.

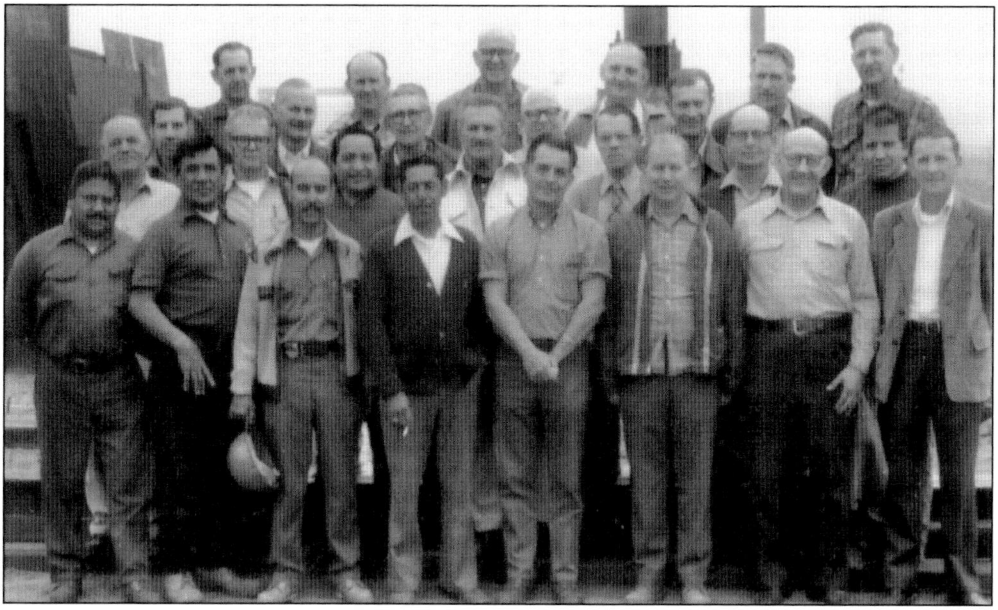

These "fab" workers, from left to right, are (first row) "Potato Bug," Louie Holquin, Mike Diaz, Manuel ?, unidentified, unidentified, and Tom Agic; (second row)? Manix, Walt Martin, Al Montey, Bob Petzel, unidentified, Tex Baless, and unidentified; no one is known in the third row; and (fourth row) Victor ?, Billy Gene McKee, unidentified, unidentified, Lou Turner, Bill Galbraith, and unidentified.

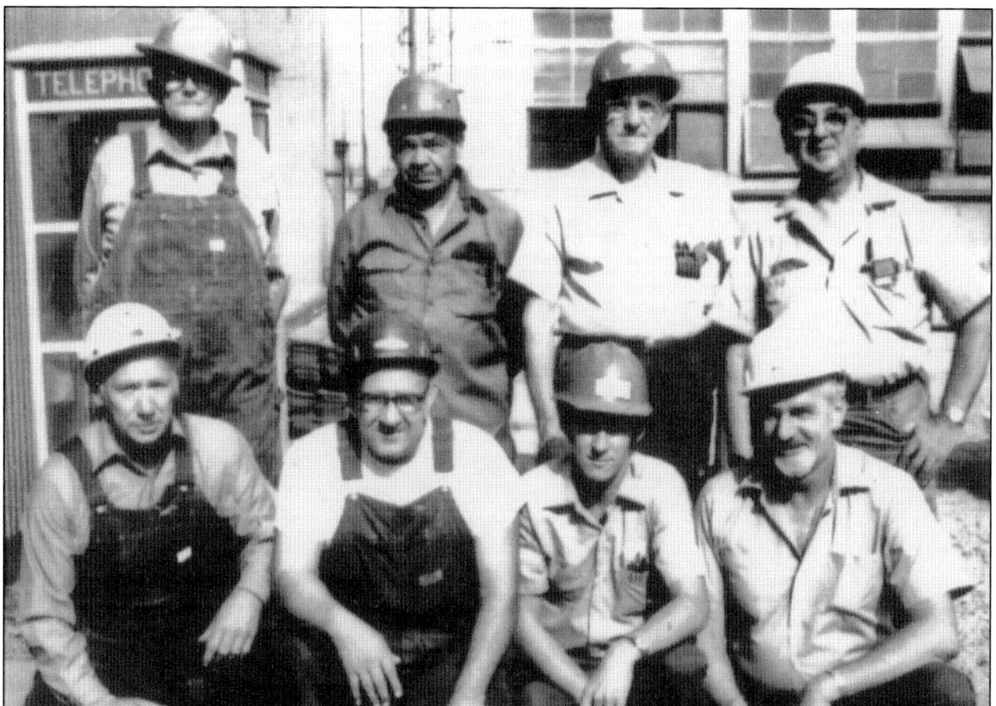

Pictured, from left to right, are (first row) Jerry, Joe, and Billy Gene McKee, and Joe ?; (second row) Walter Cook, Poncho, Thomas, and Bob S.

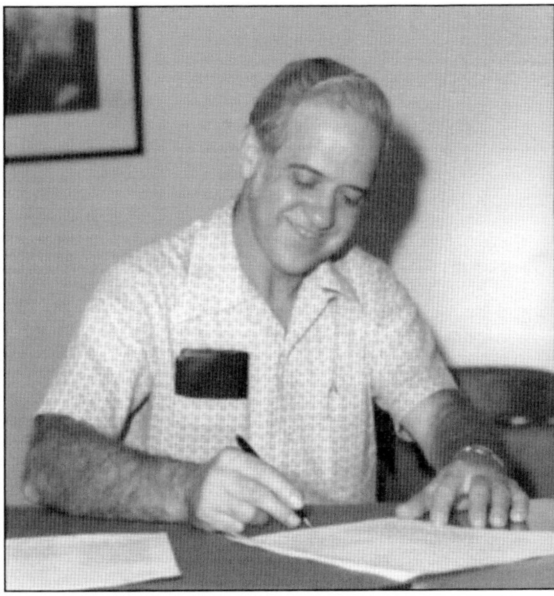

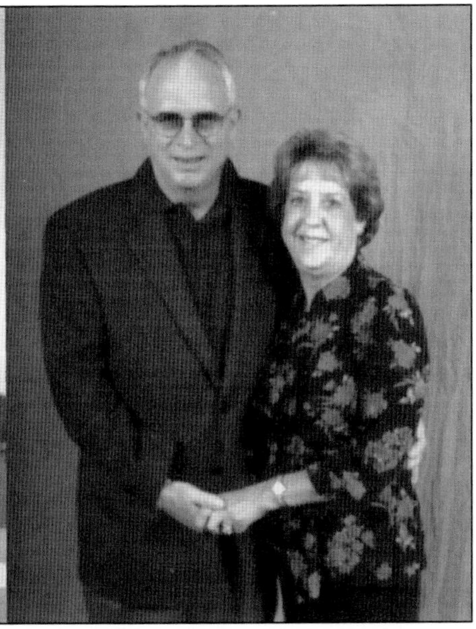

Roy Sanders worked for Kaiser for 30 years. He was from Kentucky by way of Florida. His daughter, Linda, worked as a steno-clerk for 9 years and married Steven Ipson, who worked for 17 years in production planning, and then for California Steel for 20 more years. He is a native Californian.

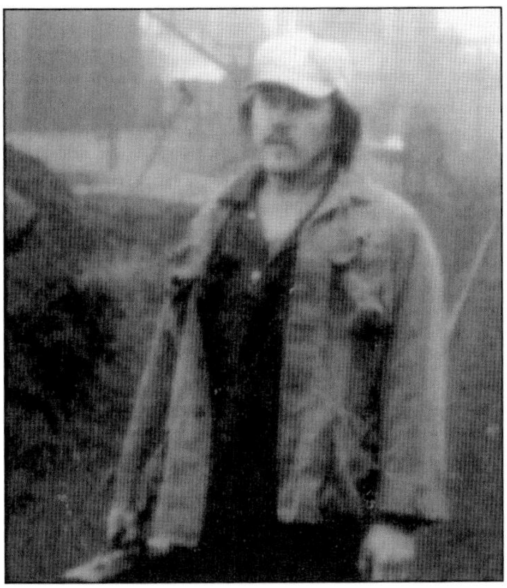

Bert Bene worked at Kaiser for approximately 18 years at the tin mill's open hearth.

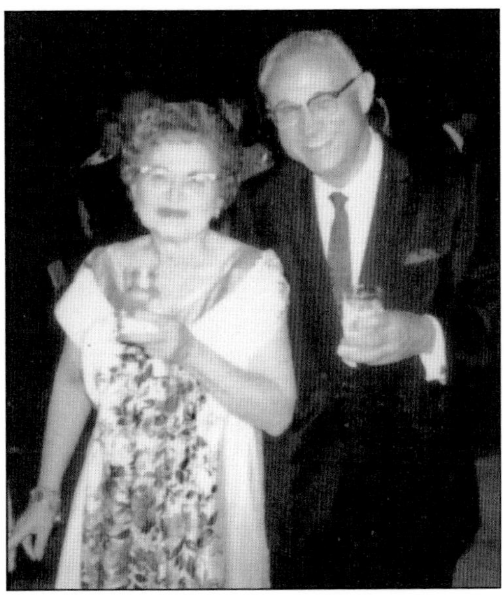

Ray Robertson worked for Kaiser for 32 years as director of public relations. This photograph depicts Ray and his wife, Carol, at a company social.

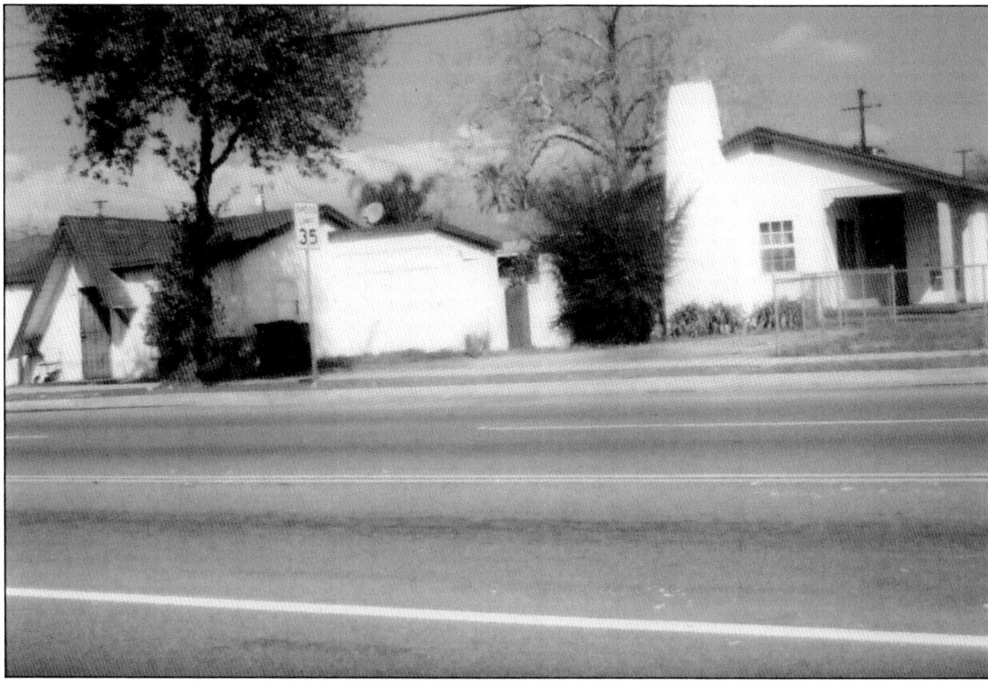

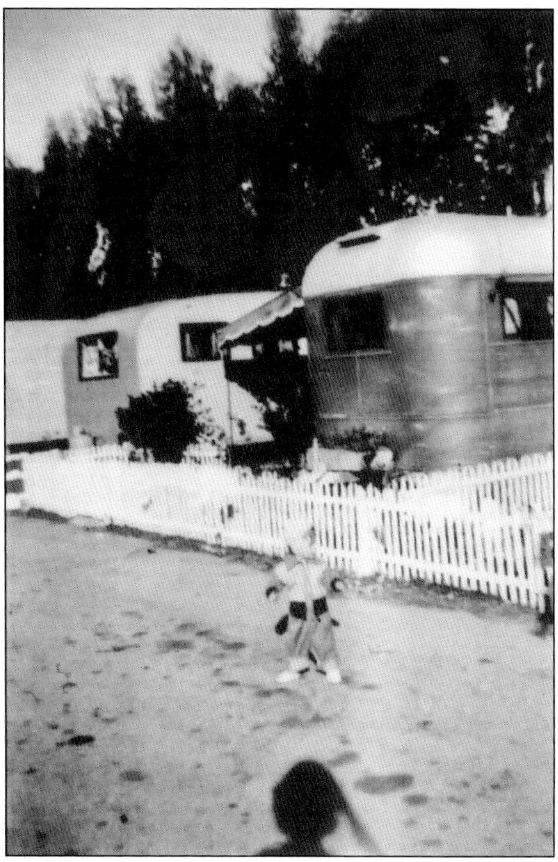

The "guest CEO" housing for the mill was located on Miller and Palmetto Avenues. The Kaiser guests used to stay here when visiting the plant. Henry J. Kaiser had a house on the mill grounds, which was moved to near the current site of Kaiser Hospital. The building still exists. (Courtesy of author.)

In 1948, the Kaiser trailer park for employees was located on San Bernardino Avenue near the plant. The little boy is Larry Husted, who is featured elsewhere in this book. (Courtesy of Larry Husted.)

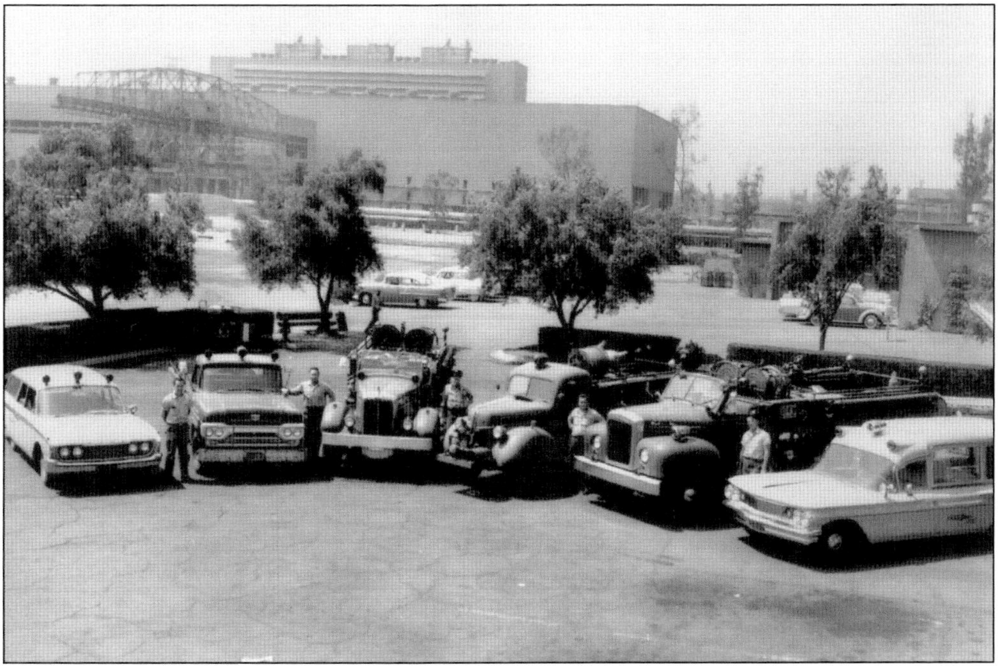

An operation the size of Kaiser Steel had to function as if it were a heavily industrialized city unto itself. With all of the fire and chemicals involved in steelmaking, the plant required its own fire-control operations. These two photographs are of the Kaiser Fire Department. One of Kaiser's firemen was Al Chavez, whose widow, Lois, is a member of the board of the new Fontana Heritage Museum Association. (Courtesy of Kaiser High School Library.)

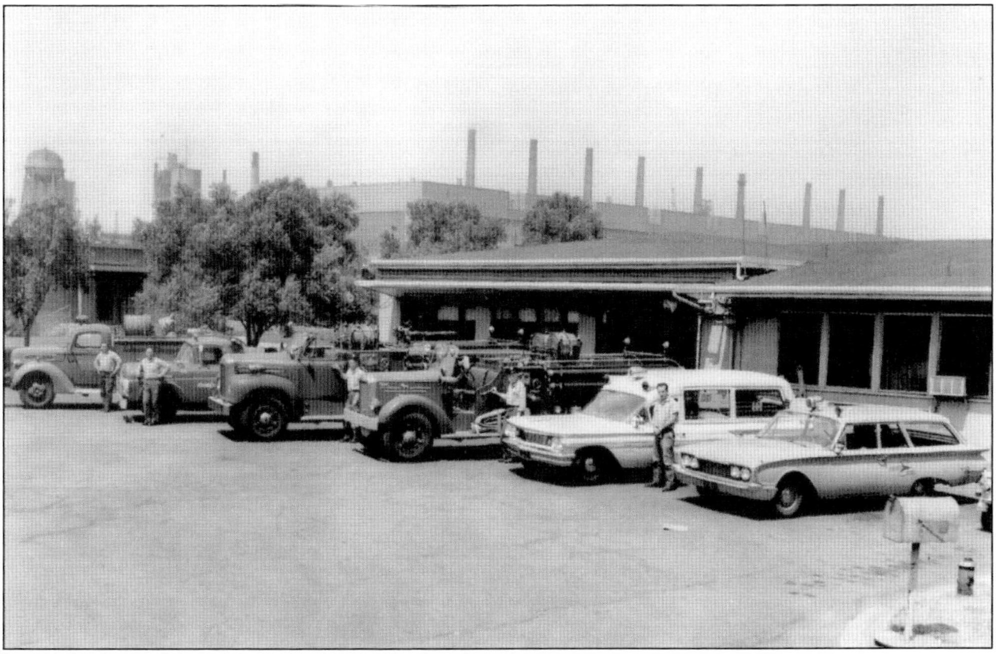

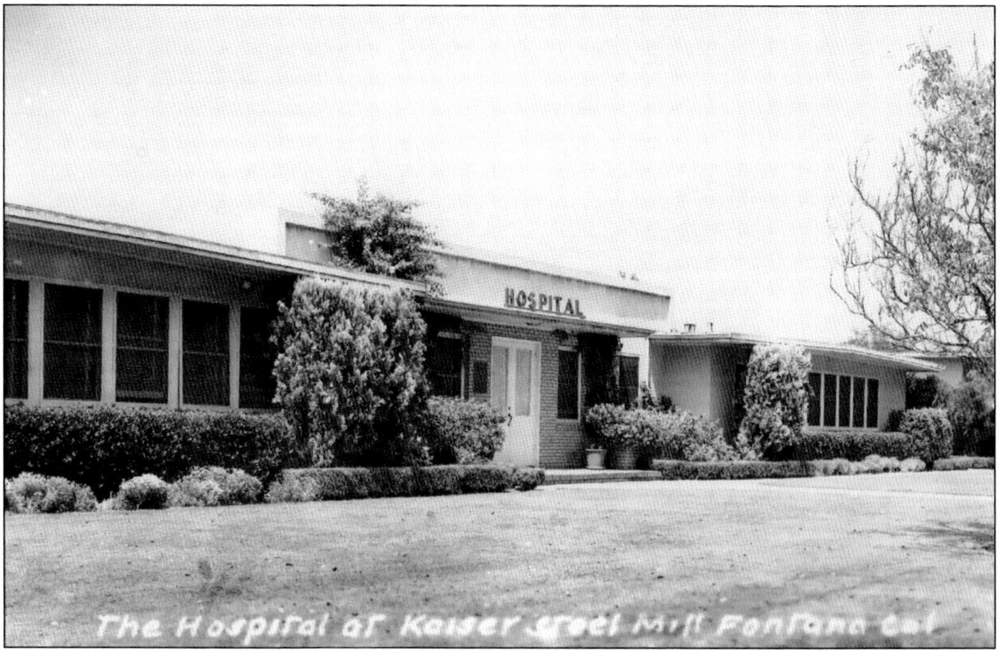

These images show Kaiser Hospital on its original Kaiser Steel Plant and Mill site, north of the administration building. Many of the employees' children were born here. In a lot of instances, the father would be called in from work in the mill to be there for the delivery of his son or daughter.

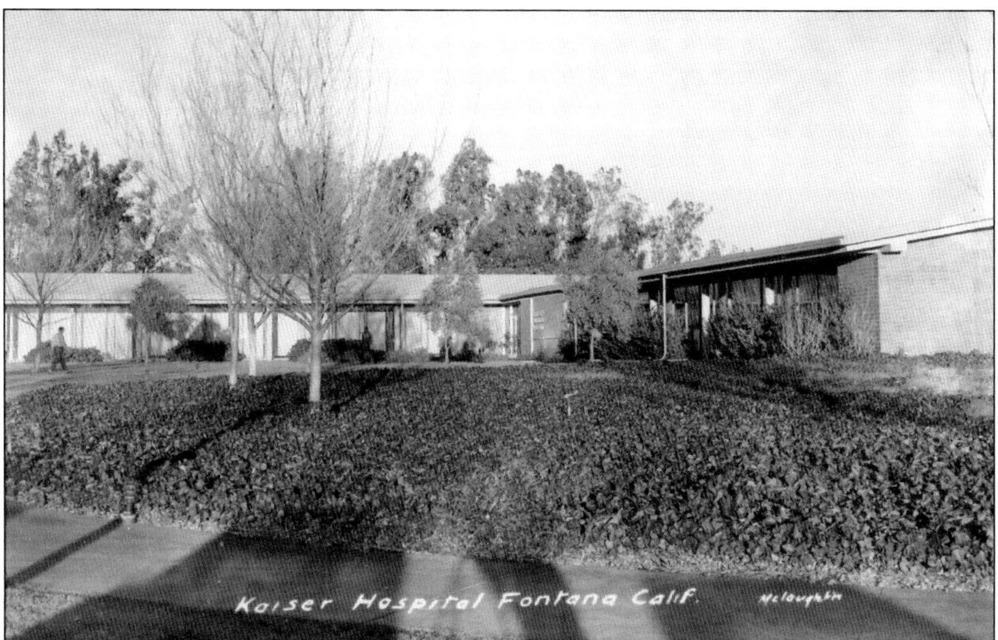

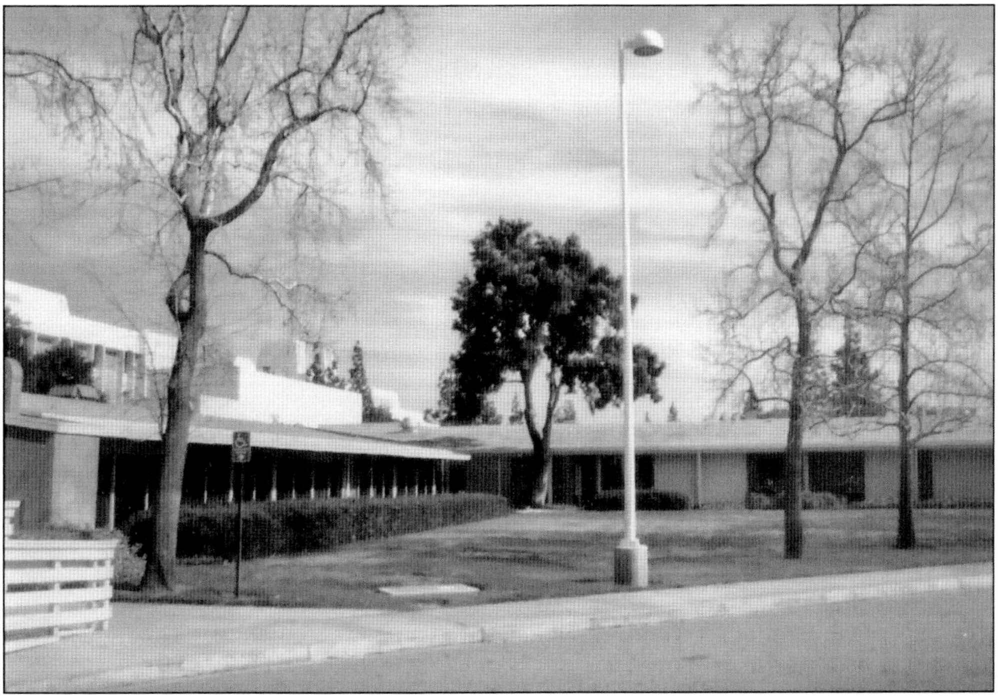

This was the same Kaiser Hospital, but at a new location, on Sierra and Marygold Avenues. The original medical center was becoming too small to serve the large employee population and its families. But when the new hospital became too small after a short time, it was decided to move the original hospital to a new location as an add-on, where it stands today.

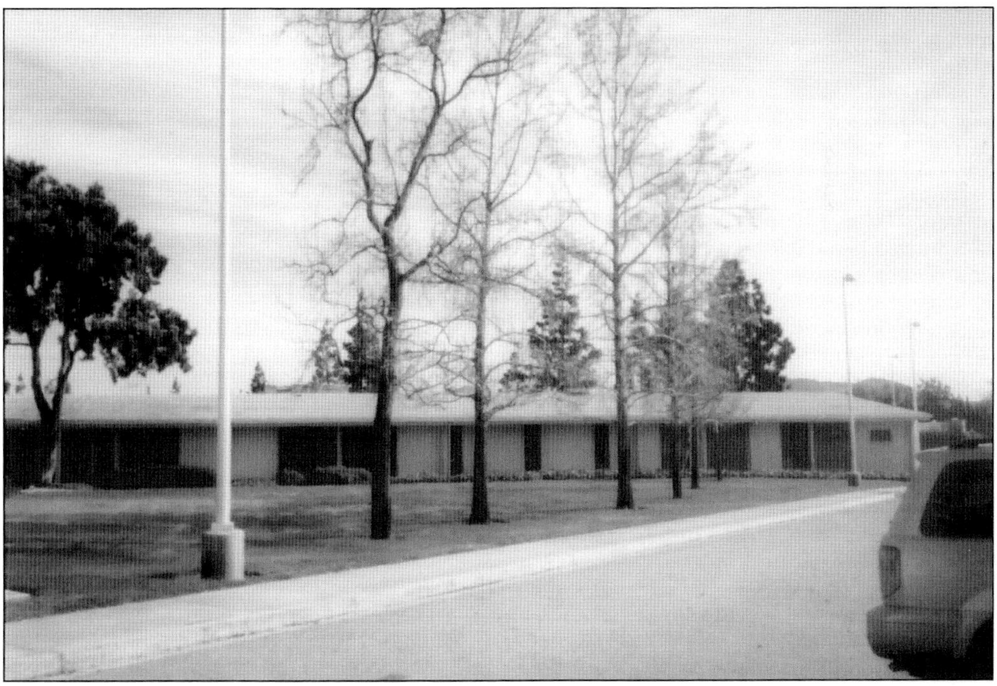

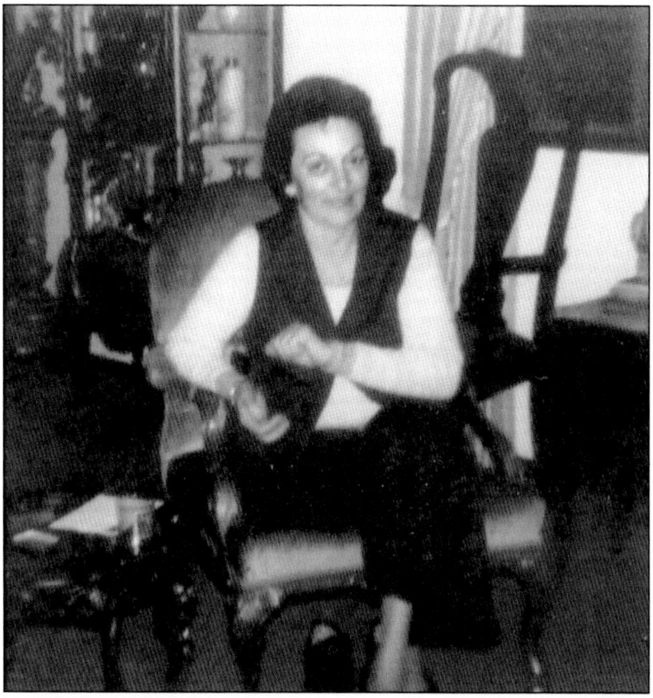

Sophia Momeilovich worked at the old Kaiser Hospital for many years. (Courtesy of Jackie Ficarotta.)

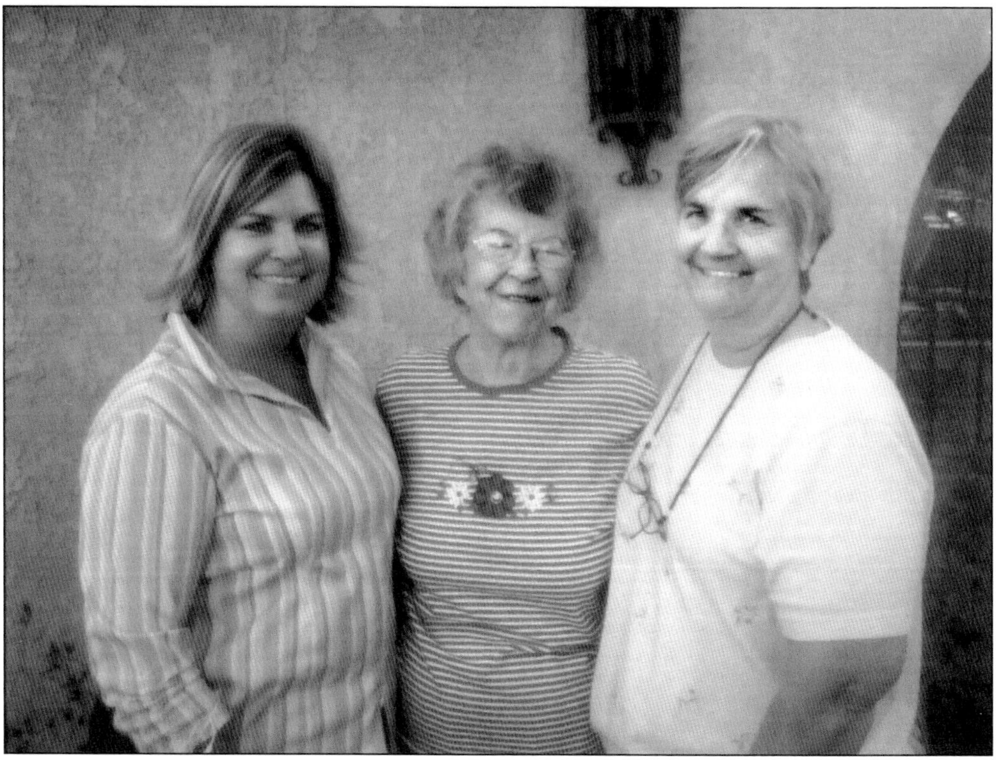

Health care spans across generations here as these three ladies have all worked for Kaiser Hospital. Ruth Covert (center) is the wife of Timon "Curly" Covert. At right is her daughter, Ruthetta Brandt. The woman at left is Ruthetta's daughter, Jeanine Davis. (Courtesy of the author.)

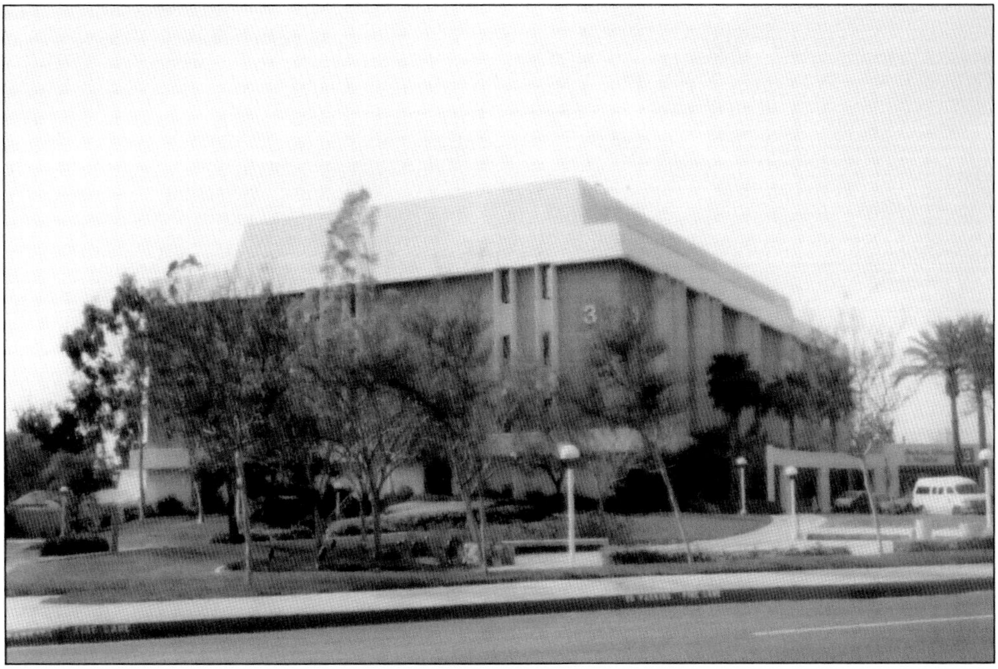

These two images are of Kaiser Permanente Hospital, the largest HMO in the world. It started out as a clinic operated by Dr. Sidney A. Garfield at Desert Center, near Eagle Mountain, during the Great Depression, and grew through the medical needs at Edgar Kaiser's facilities at Grand Coulee Dam in the state of Washington, shipyards at Richmond, California, and at Fontana and elsewhere into the giant that it is today. (Courtesy of the author.)

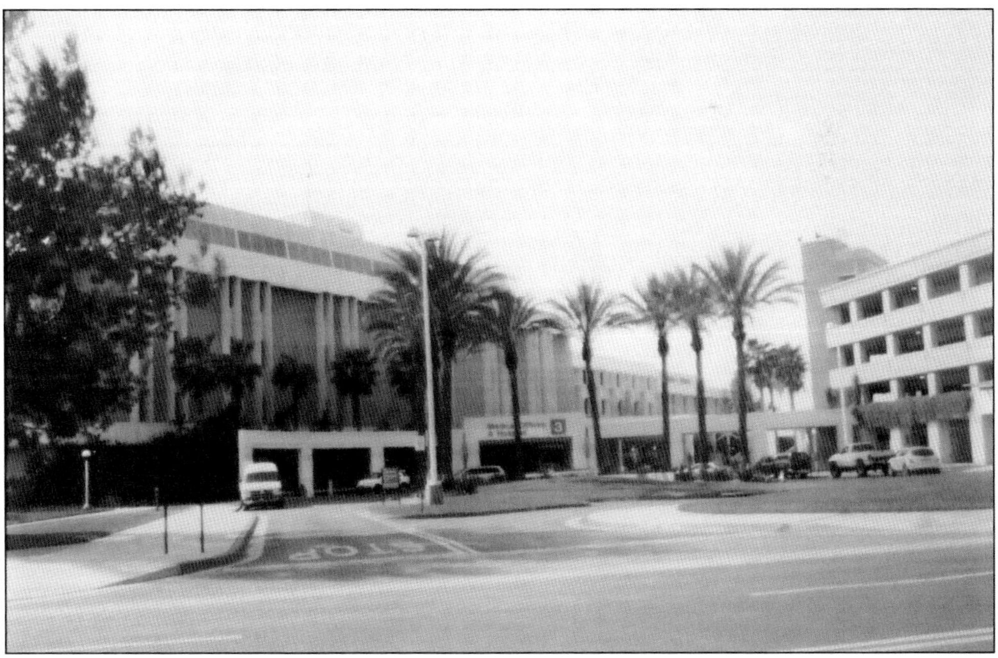

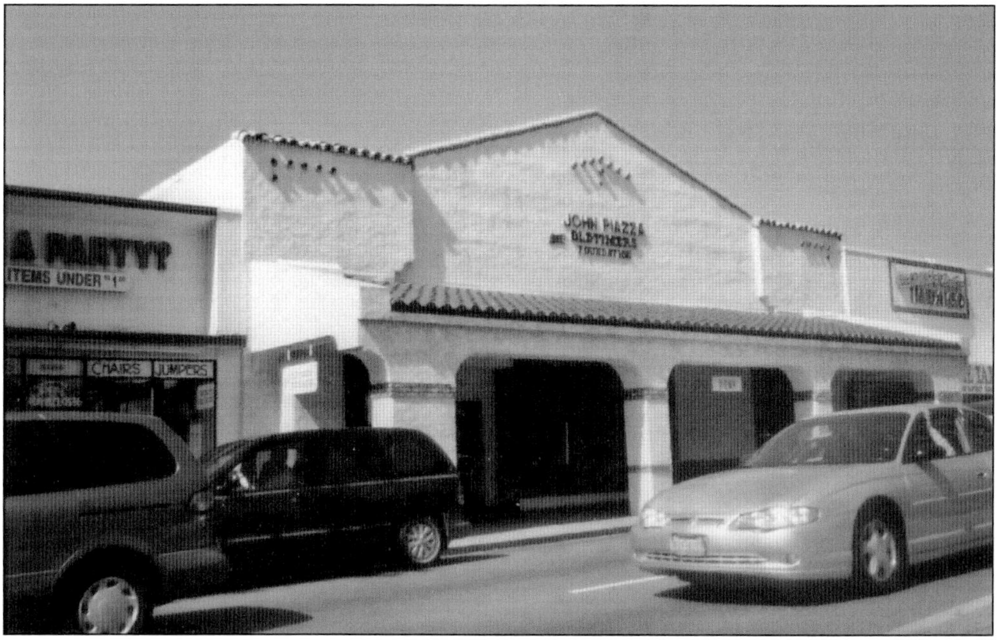

This is the John Piazza Oldtimers Foundation office on Sierra Avenue, below Arrow Highway on the west side of the street. It used to be a Safeway store. (Courtesy of the author.)

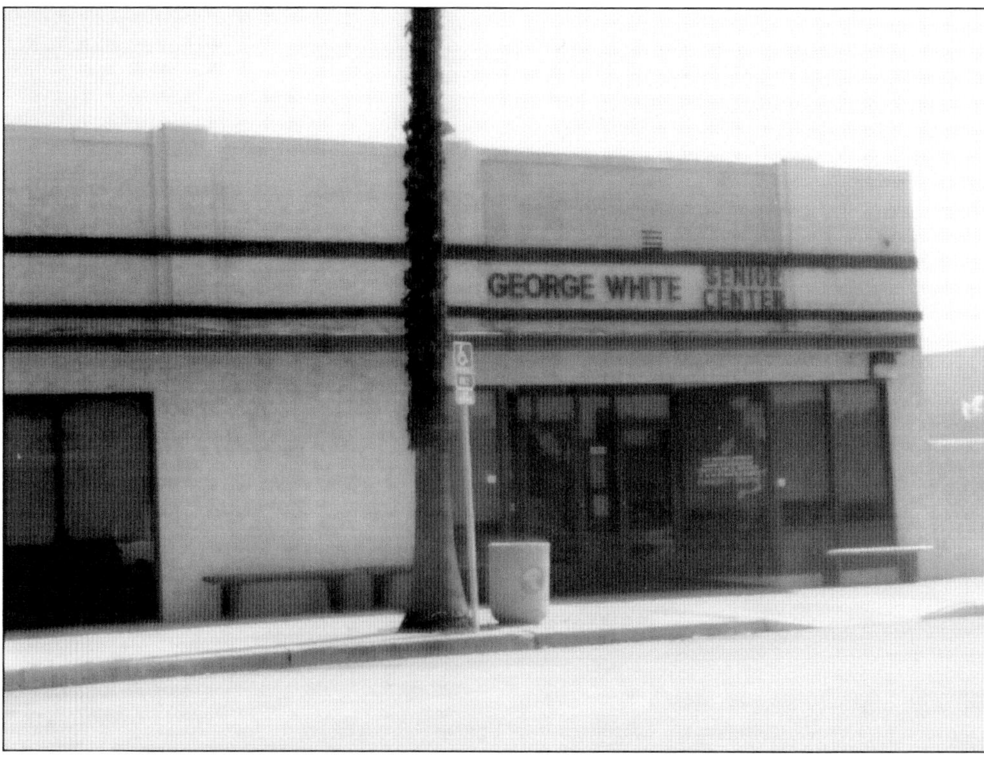

The George White Senior Center is located on Nuevo Avenue on the east side of the street just above Valencia Avenue. Lunches are served here to seniors. (Courtesy of the author.)

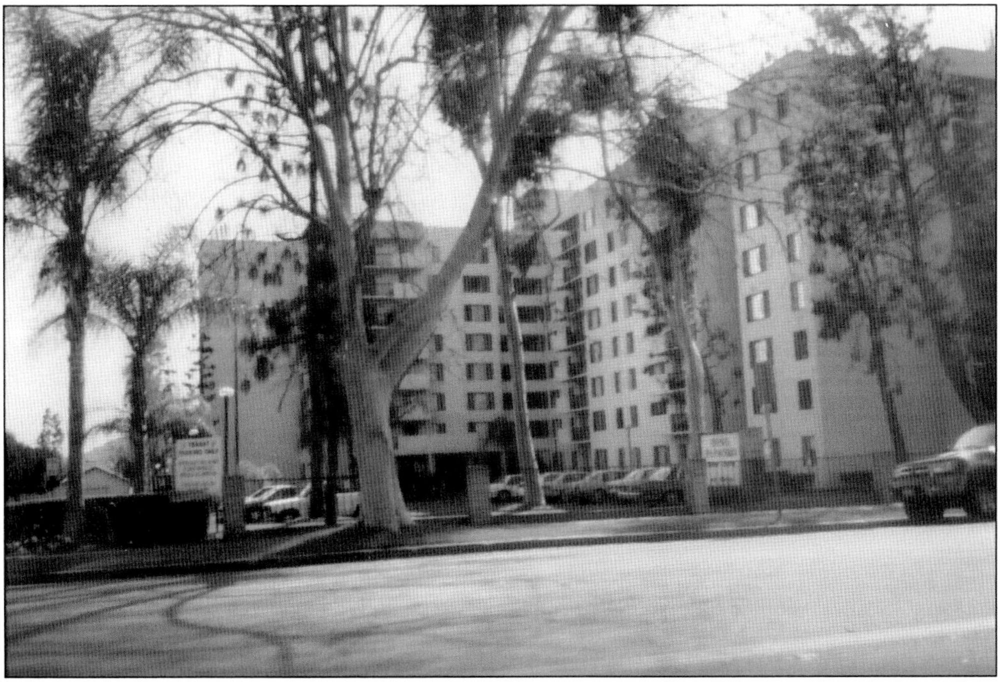

The Marygold Senior Highrise, located on the southeast corner of Marygold and Juniper Avenues, is eight stories tall and presently the highest in the city of Fontana. It was named after Dino Papavero. (Courtesy of the author.)

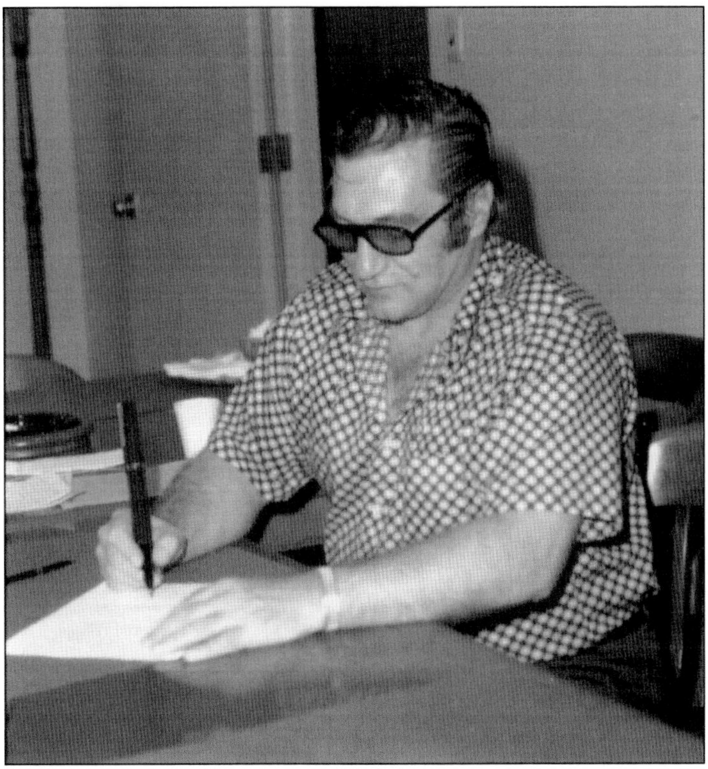

This photograph shows one of the great friends of the Fontana workingman, Dino Papavero of the United Steelworkers of America. (Courtesy of the Timon Covert family.)

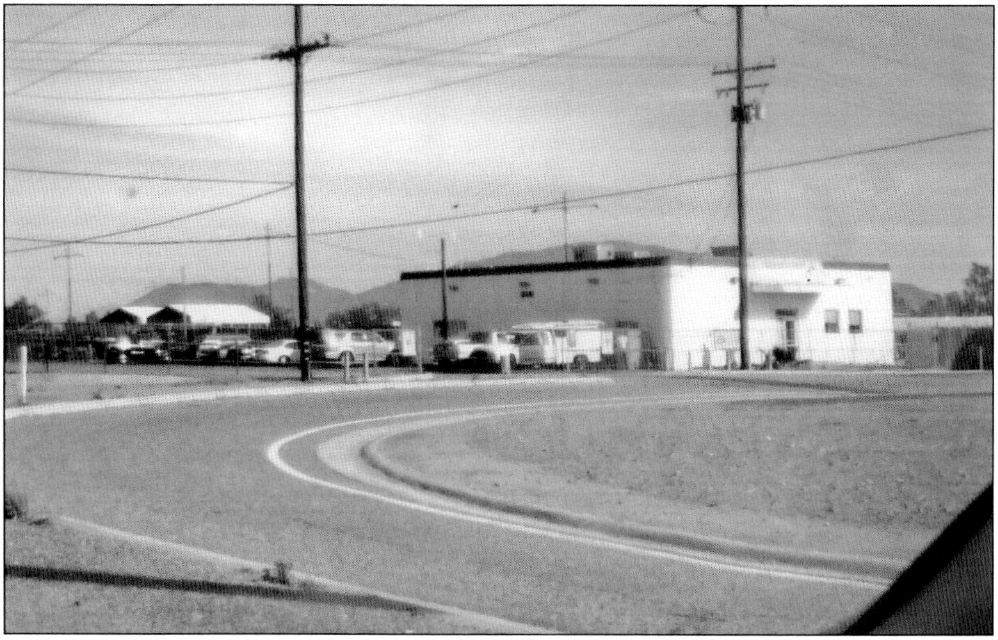

This image depicts the old Union Hall on San Bernardino Avenue south of the mill. (Courtesy of the author.)

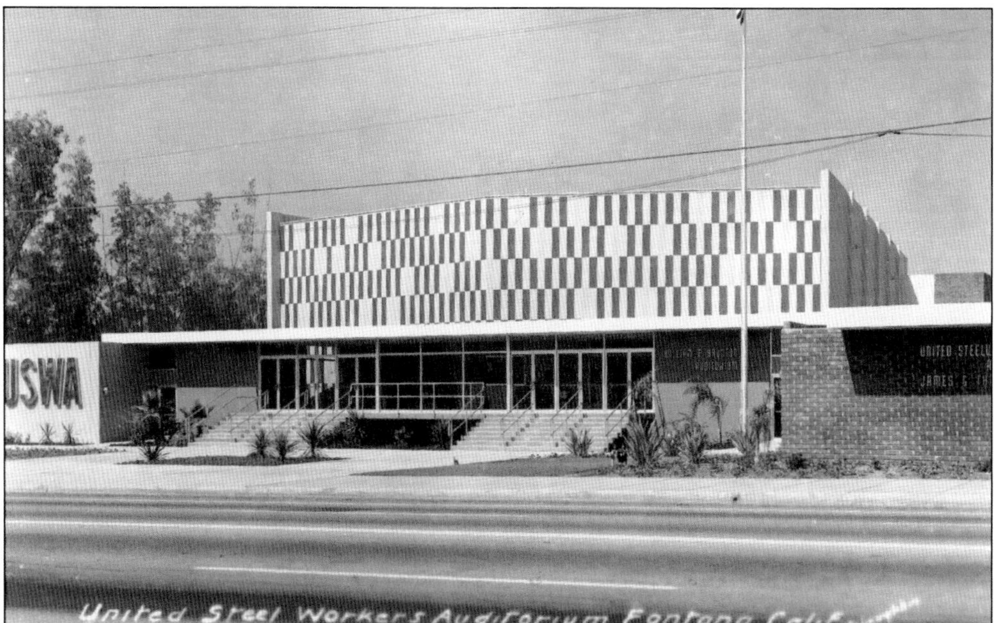

The new Union Hall and Auditorium on Sierra Avenue was home to the United Steelworkers of America. The building is now the civic auditorium for the city of Fontana and is used for the parks and recreation offices. It may be earmarked for demolition. The museum now has the original plate for printing this photograph.

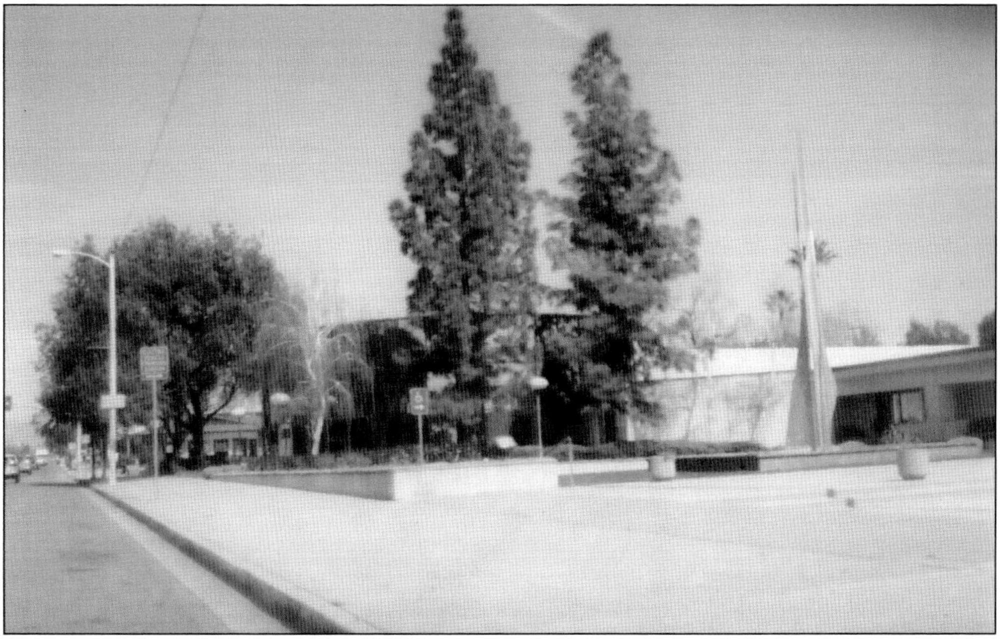

Fontana City Hall on Sierra and Upland Avenues is located on the southeast corner. It was built in 1961 and added to over the years. The soaring steel fountain that graces its front was donated by Kaiser Steel and arrived in 1962. Kaiser commissioned it from internationally known sculptor Francois Stahly, who designed the fountain at the Seattle World's Fair. (Courtesy of the author.)

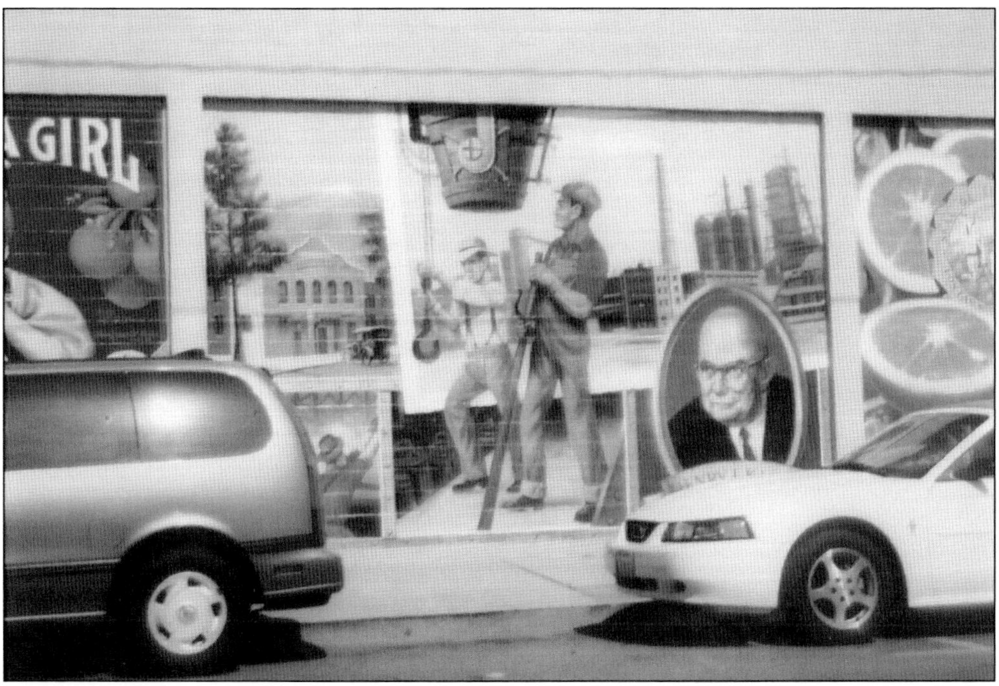

Painted in 2005, the four-section mural on the south side of the building on Sierra and Valencia Avenues contains a portion dedicated to Fontana's Kaiser Steel Mill and the steel industry.

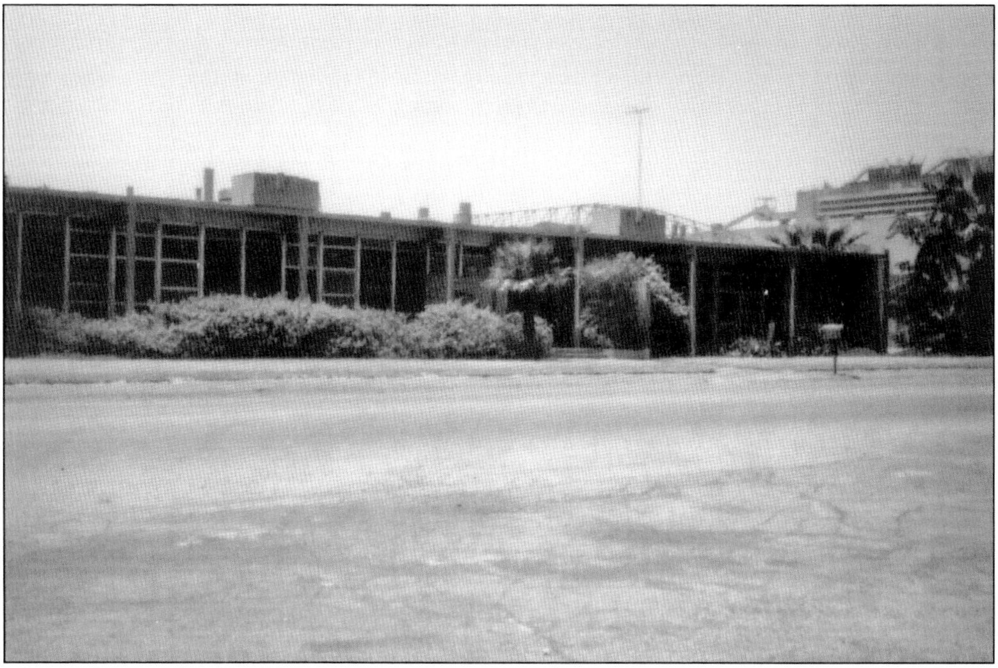

The in-plant medical building is pictured here after the closing.

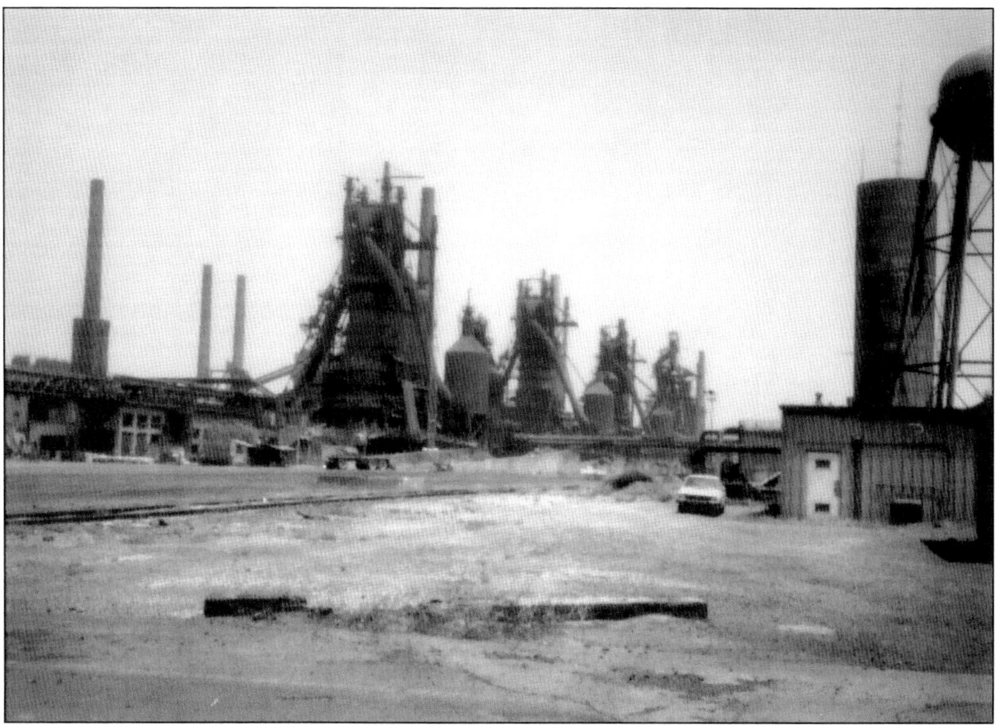

These are the formerly mighty No. 1 through 4 blast furnaces, pictured after the closing.

Five

Views of the Closed Kaiser Plant

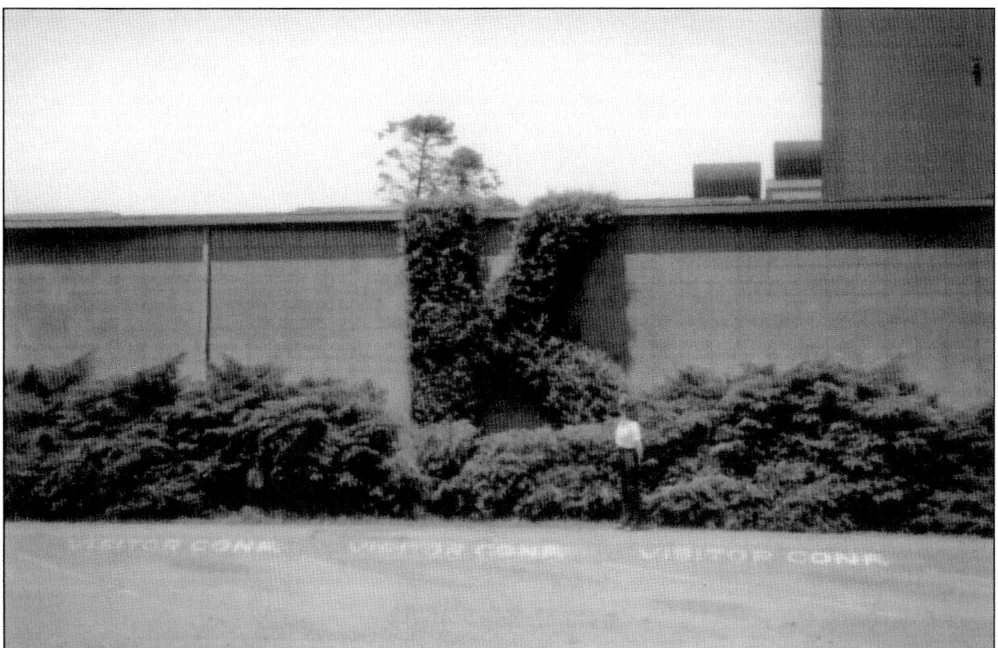

This image shows the Kaiser administration building in 1983 after the mill closed down. Jonnie Long remained at the facility for two years after to finish the work in the claims department. She is standing in front of the hedge at the "K" logo. (Courtesy of Jonnie Long.)

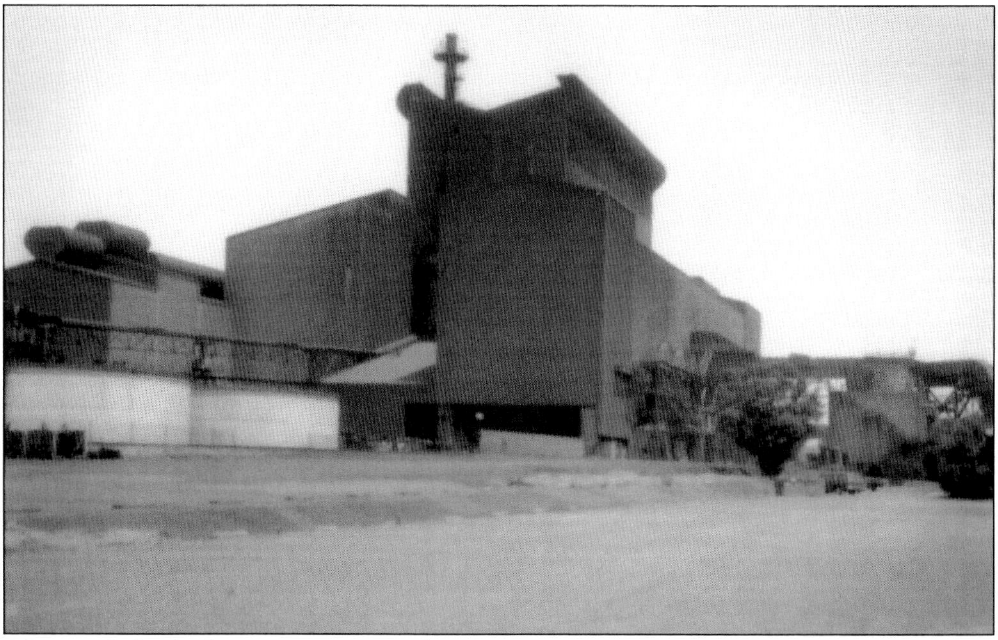

BOP stands for the basic oxygen-process building. (Courtesy of Jonnie Long.)

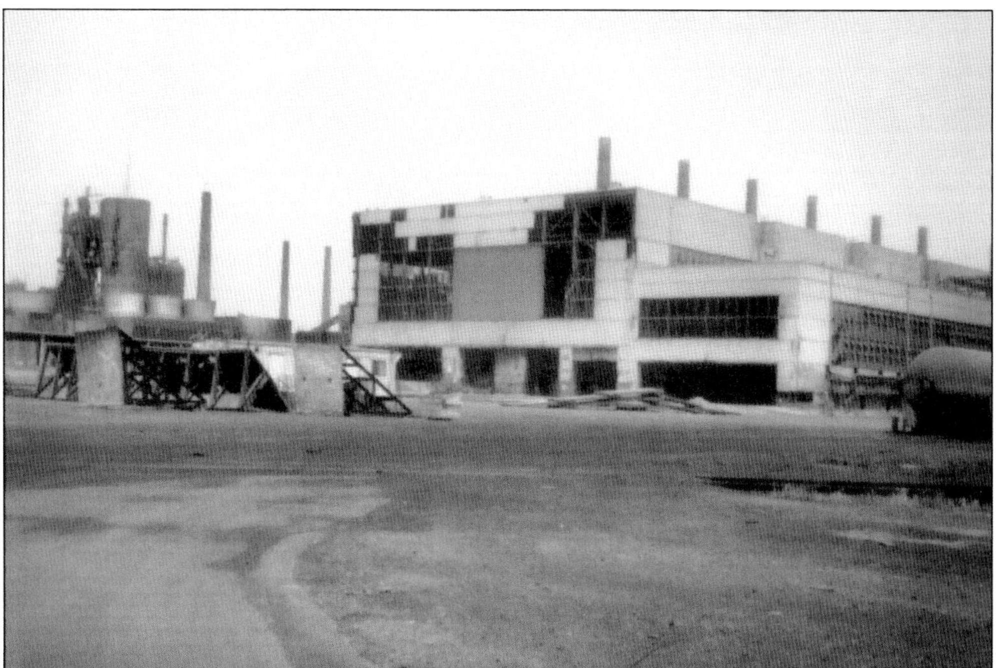

This is the west end of the open hearth, south of the blast furnaces. (Courtesy of Jonnie Long.)

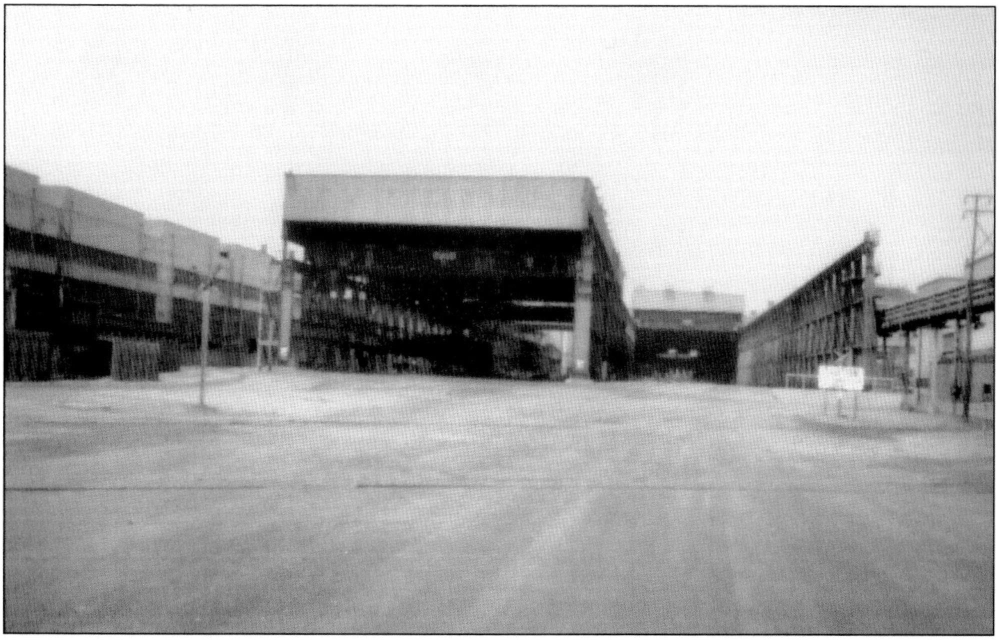

The workers' locker room is pictured here at left and the slab yard at right. (Courtesy of Jonnie Long.)

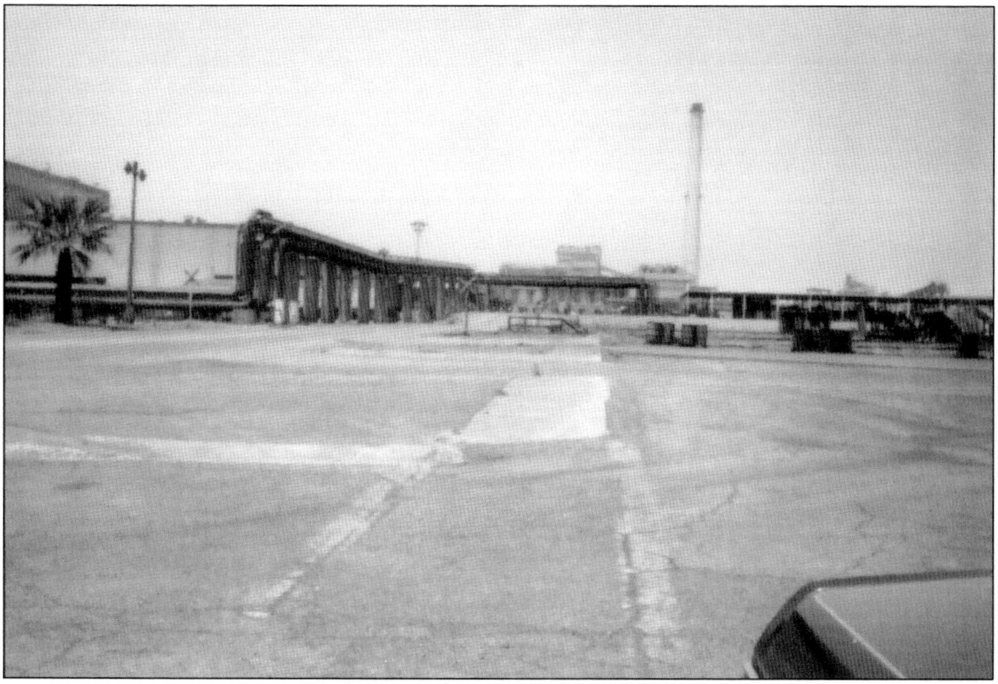

This image looks west, north of the BOP. (Courtesy of Jonnie Long.)

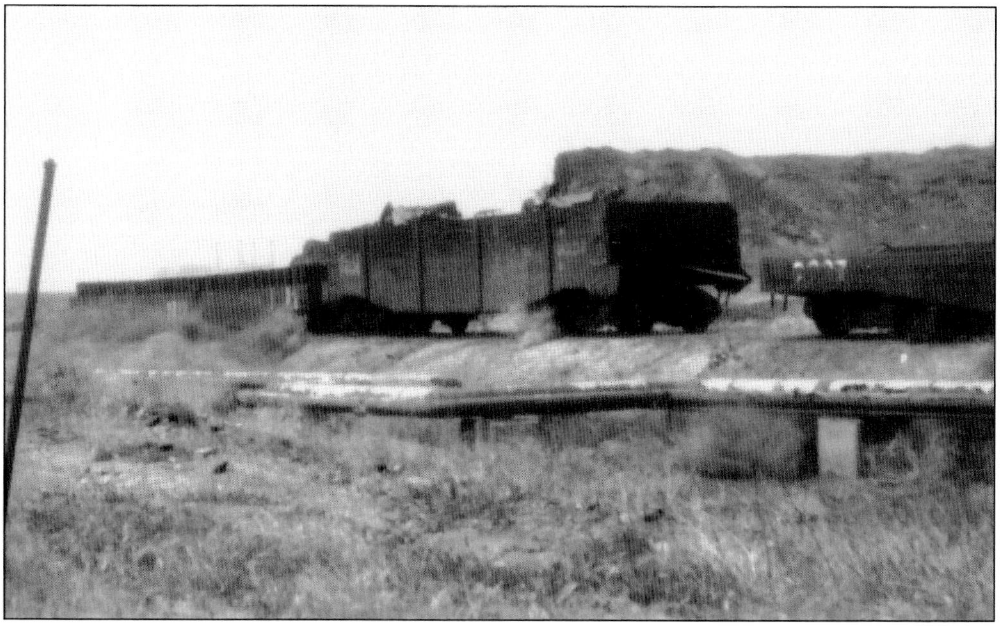

The raw materials stockpiling area is pictured here with old train cars. (Courtesy of Jonnie Long.)

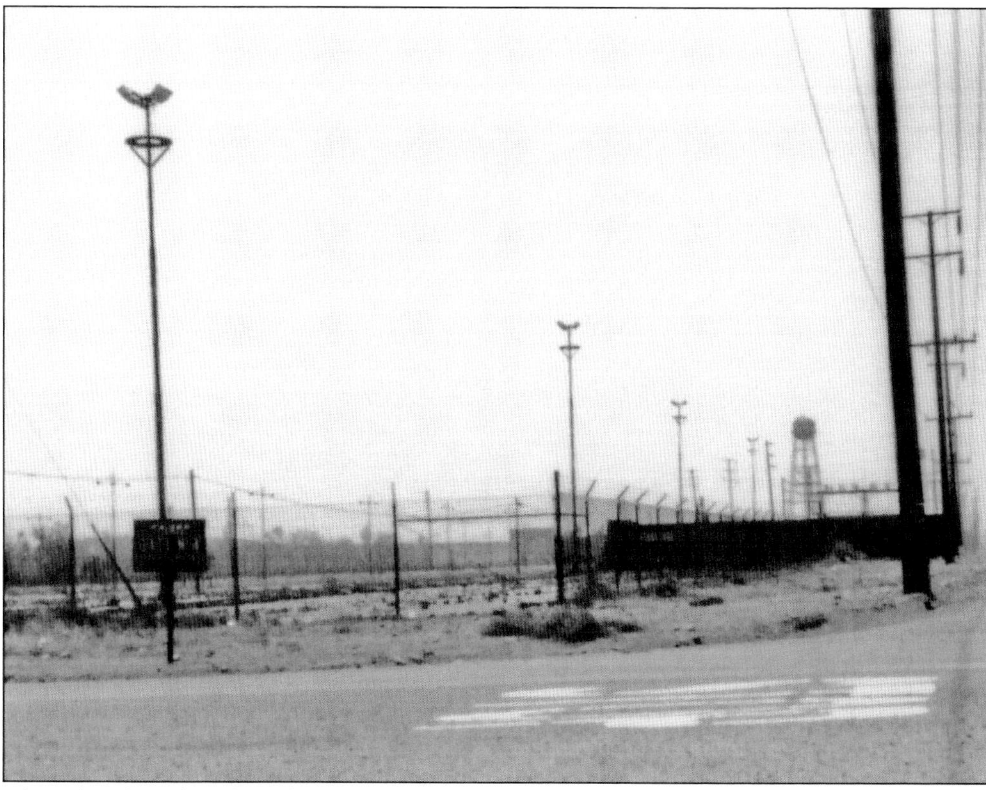

This abandoned parking lot was once filled with workers' cars at Gate 4. (Courtesy of Jonnie Long.)

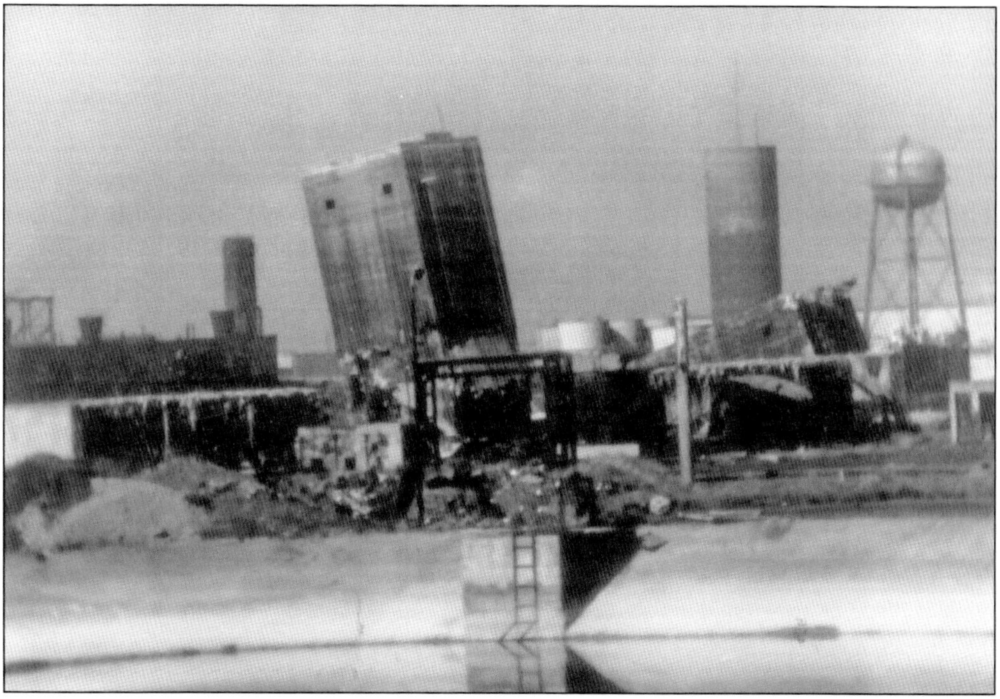

These images show the demolition of the open-hearth area to make way for the California Speedway. This area has often been used by film crews to make movies and television shows. (Both photographs courtesy of Bert Bene.)

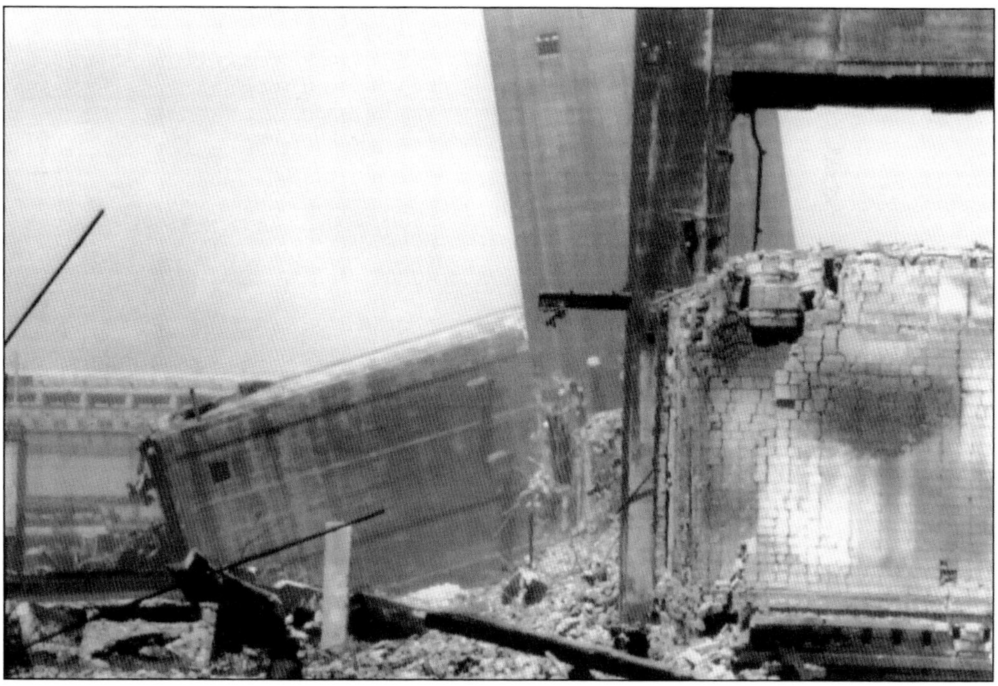

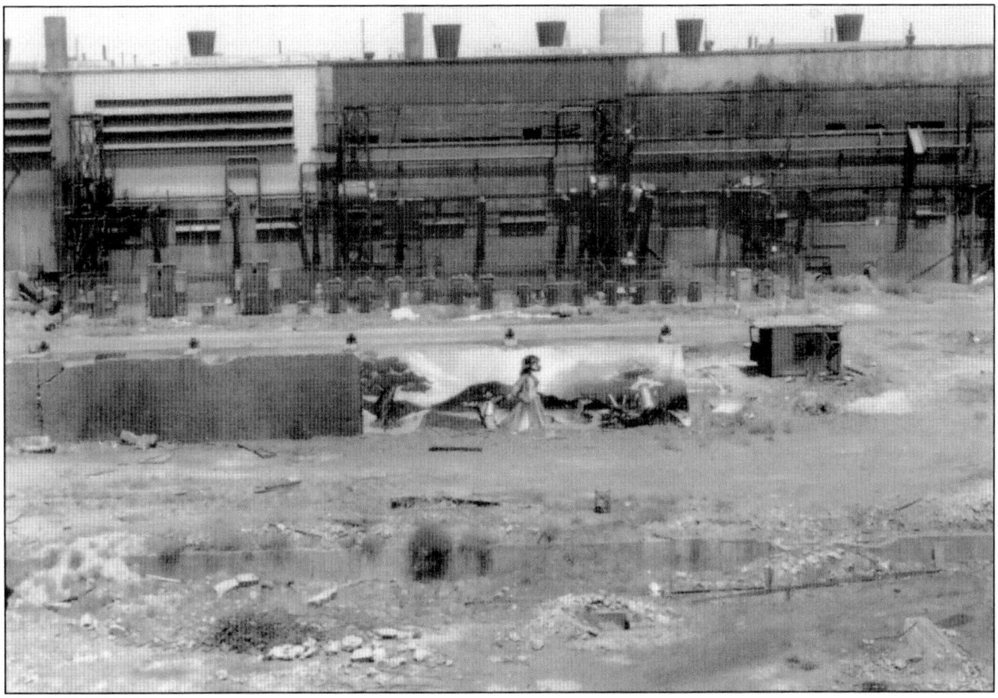

The above photograph shows the demolition, with a mural in the foreground. Below is a close-up of the mural.

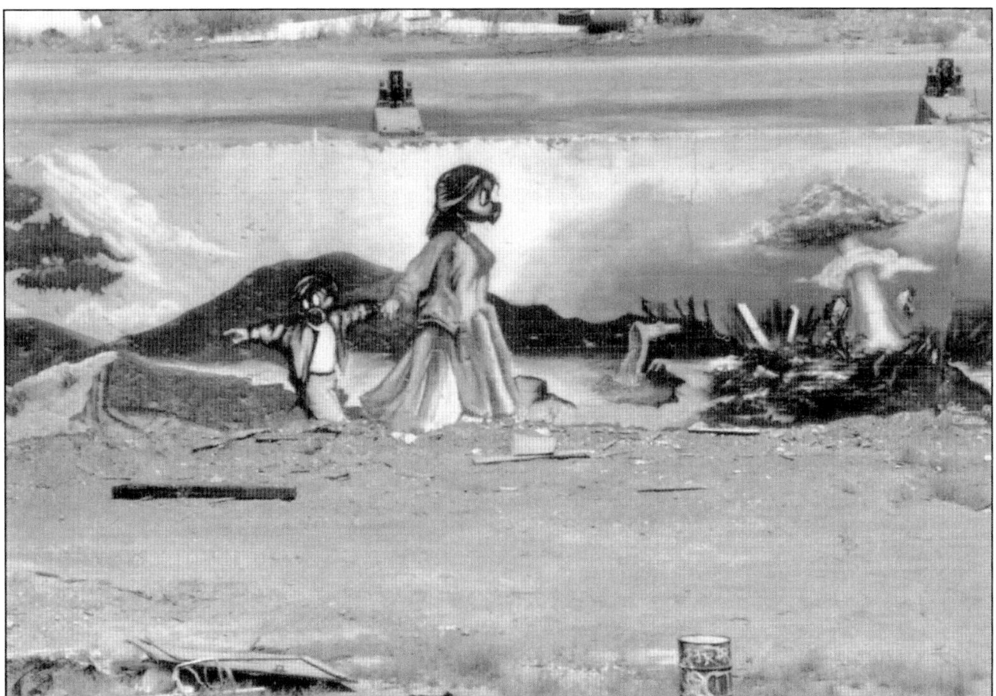

Six

CALIFORNIA STEEL INDUSTRIES

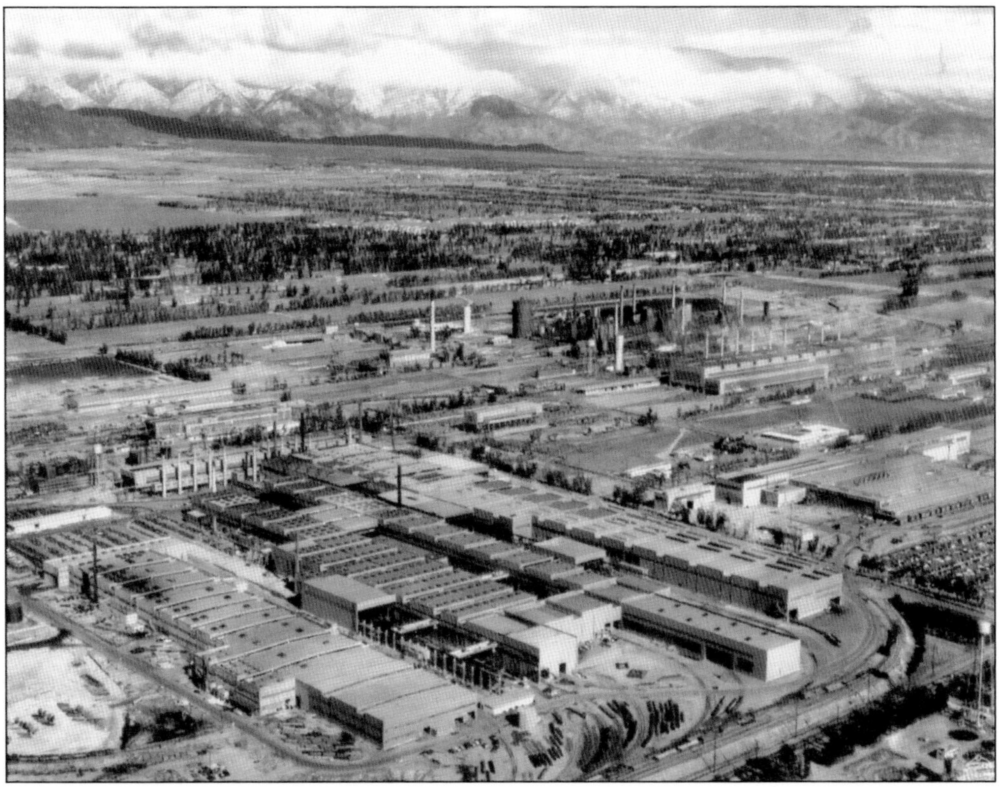

This aerial view shows the Kaiser Steel property, which is now owned by California Steel Industries (CSI). The buildings in the photograph's lower half are CSI buildings; those in the upper half were destroyed to create the California Speedway. In 1984, after the closure of Kaiser Steel's facilities, the need to establish a viable West Coast steel source in the United States brought together two of the world's biggest steelmakers, Japan's JFE Steel Corporation and Brazil's Companhia Vale de Rio Doce, the Rio de Janeiro-based, world's largest iron-ore producer and exporter.

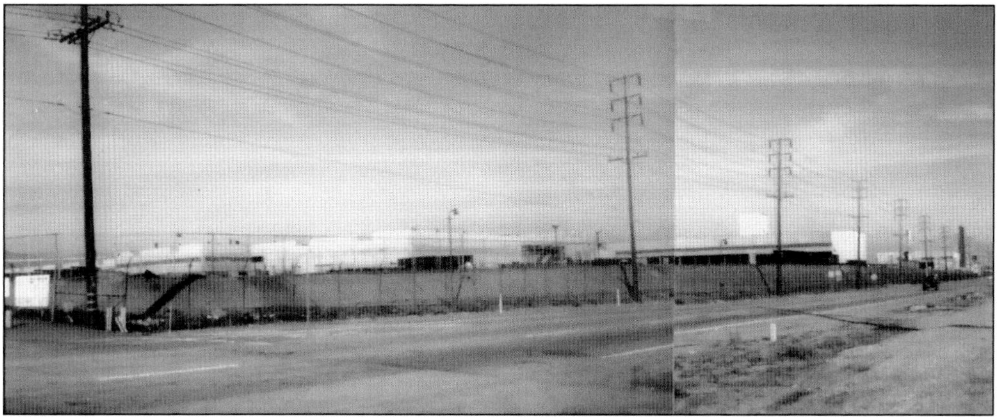

This composite panorama of the CSI buildings is from San Bernardino Avenue. The CSI plant includes an 86-inch hot-strip rolling mill, hot rolled-strip finishing facilities (build-up, shear, and slitter lines), a 62-inch continuous pickling line, a five-stand cold-reduction mill, two hot-dip galvanizing lines, cold-rolling equipment of annealing and tempering mills, and an electric-resistance welded-pipe Mill. General plant services include maintenance, shipping and material handling, wastewater treatment, medical services, and corporate offices. Athletic and recreation facilities are available for employees and their families.

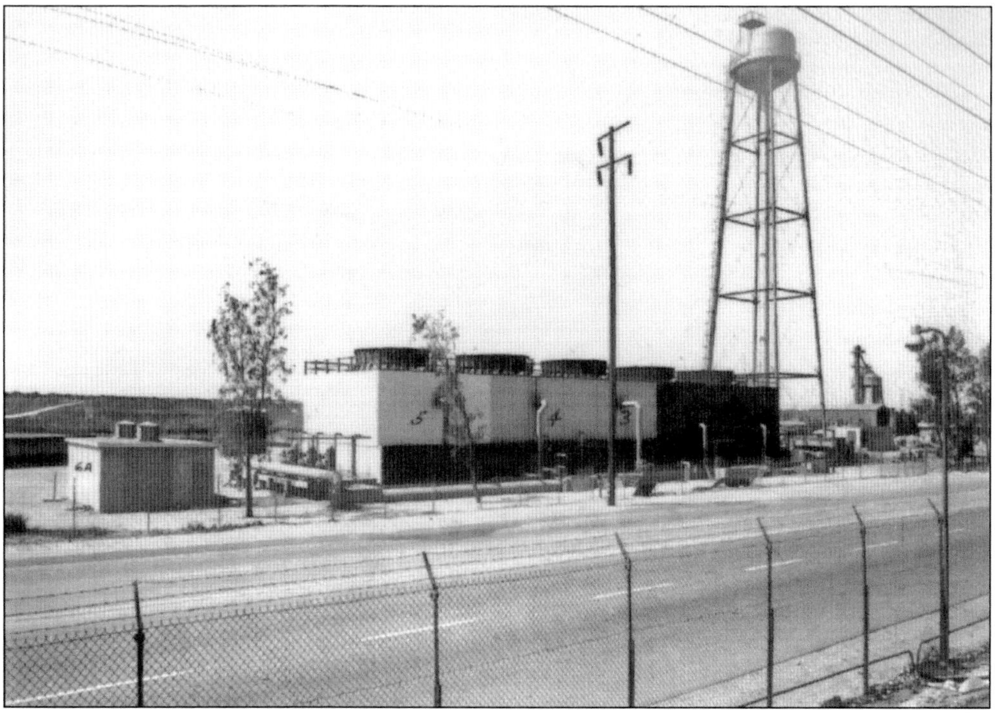

These cooling towers and processor condition the water used in the plant area. CSI has more than tripled its output and shipments since the first coils were rolled and shipped in late 1984. In 1999, CSI shipped more than 1.8 million tons of coil, sheet, and pipe products, representing sales of more than $690 million. CSI employs about 1,000 people, representing some of the finest-skilled workers in the industry. In a union-free setting, CSI cross-trains employees to perform any job on their production lines. Many former Kaiser Steel employees worked for CSI into the 21st century.

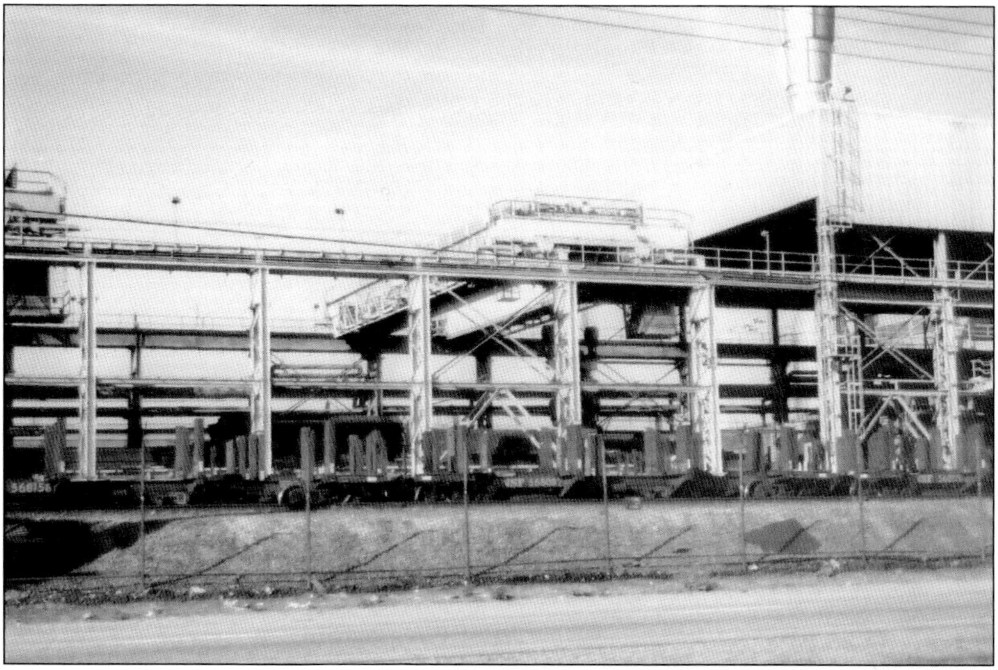

The crane in this photograph is a Heppenstal. The Santa Fe Railroad and CSI co-developed the slab cars. They were constructed specially to carry steel from Fontana to the Ports of Long Beach and Los Angeles.

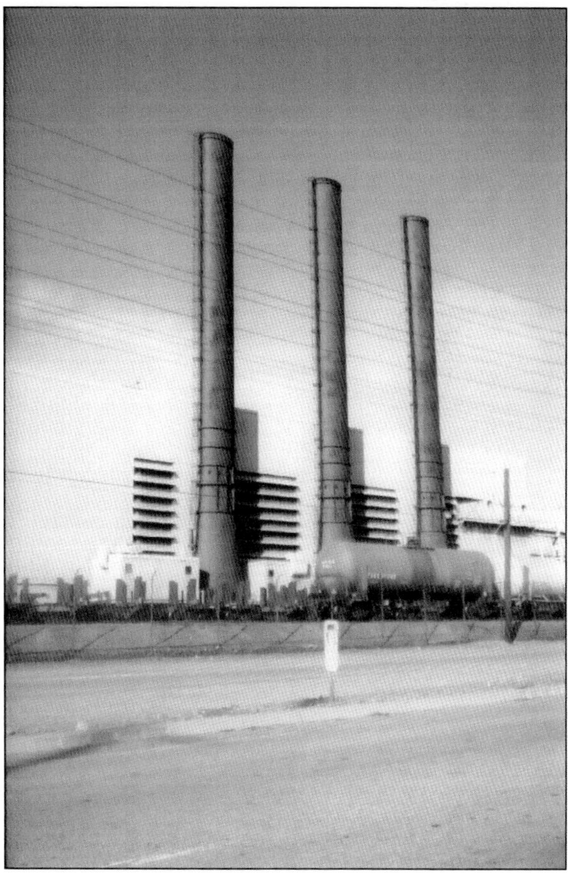

These stacks are no longer used.

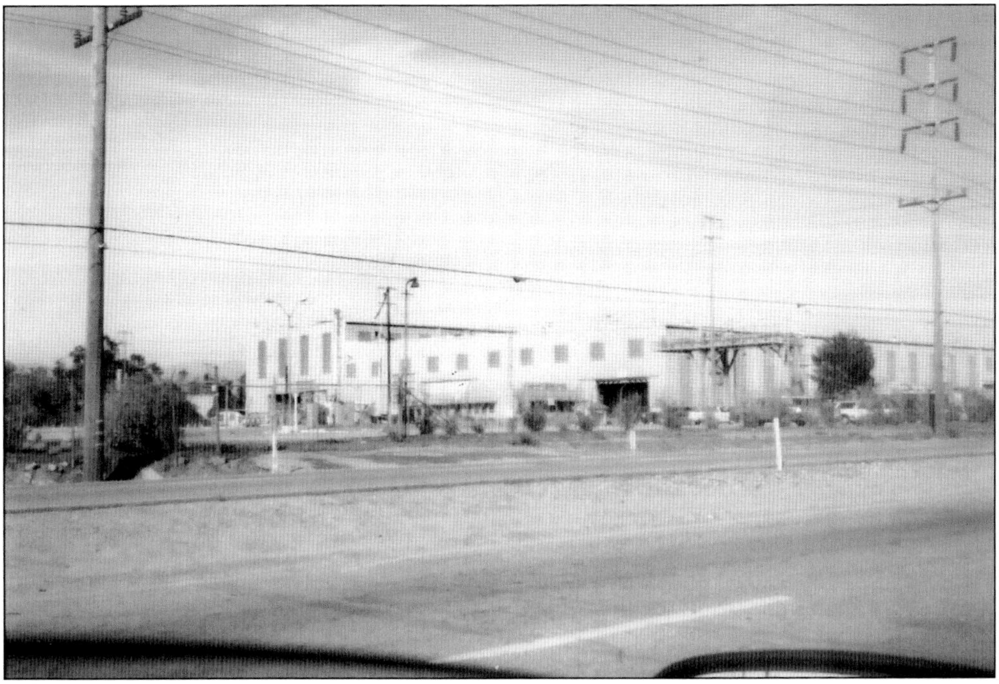

CSI's cold-roll mill is pictured here, complete with a gantry crane to load cars with coils of steel. (Courtesy of the author.)

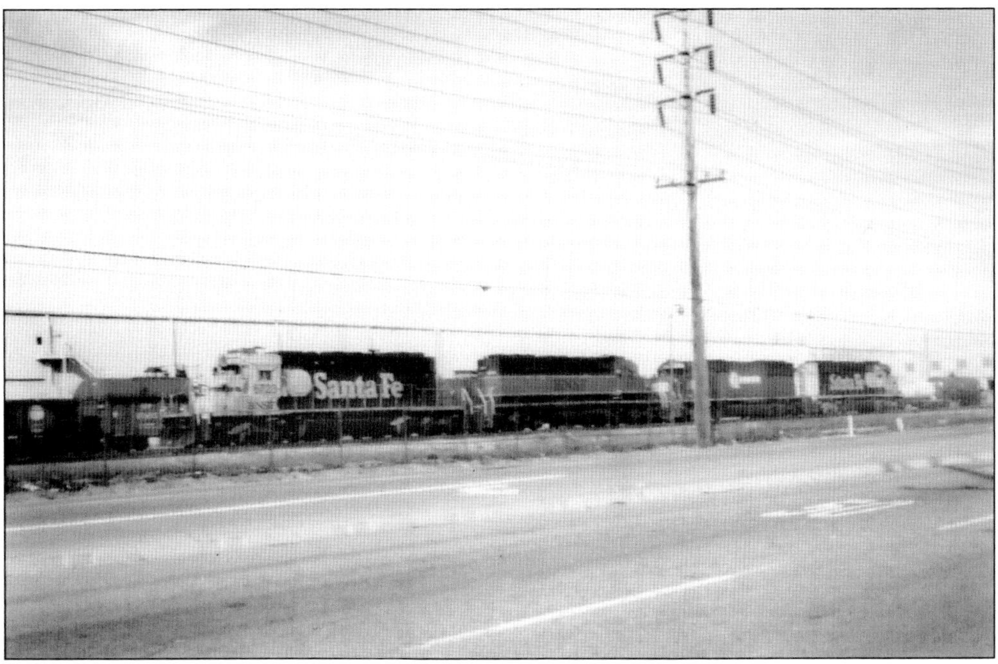

Santa Fe engines of various types are parked on CSI property until they are needed or sent somewhere else. (Courtesy of the author.)

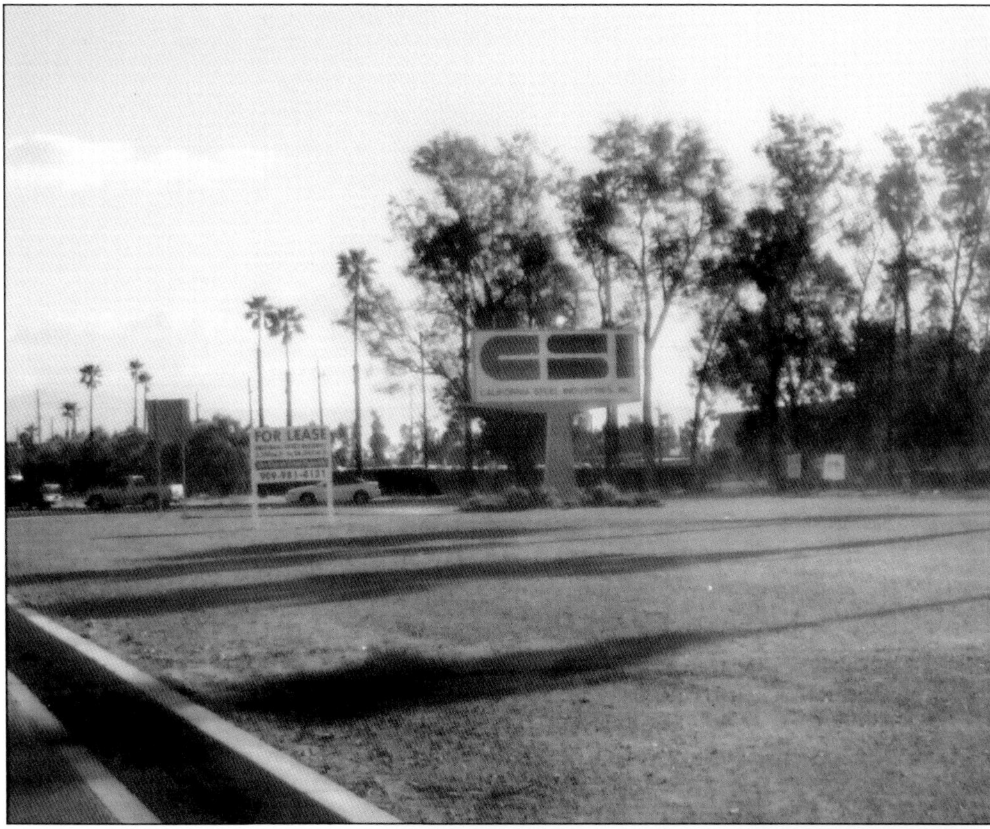

The main CSI corporate offices entrance is located off Cherry Avenue north of San Bernardino Avenue.

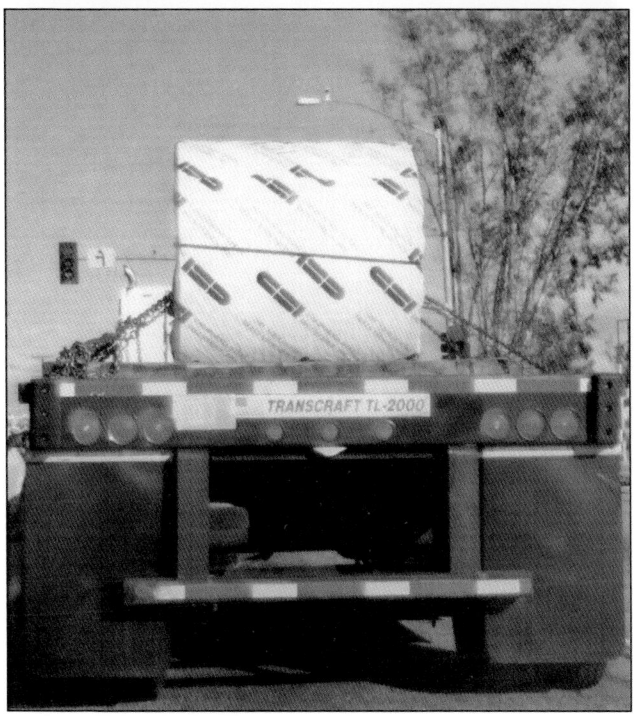

This truck is loaded with CSI-rolled steel, covered to protect it from the weather. CSI is strategically located to serve the western United States market. About 50 percent of the flat-rolled steel in this region is consumed in the boom areas of Southern California, Las Vegas, Nevada, and Phoenix, Arizona.

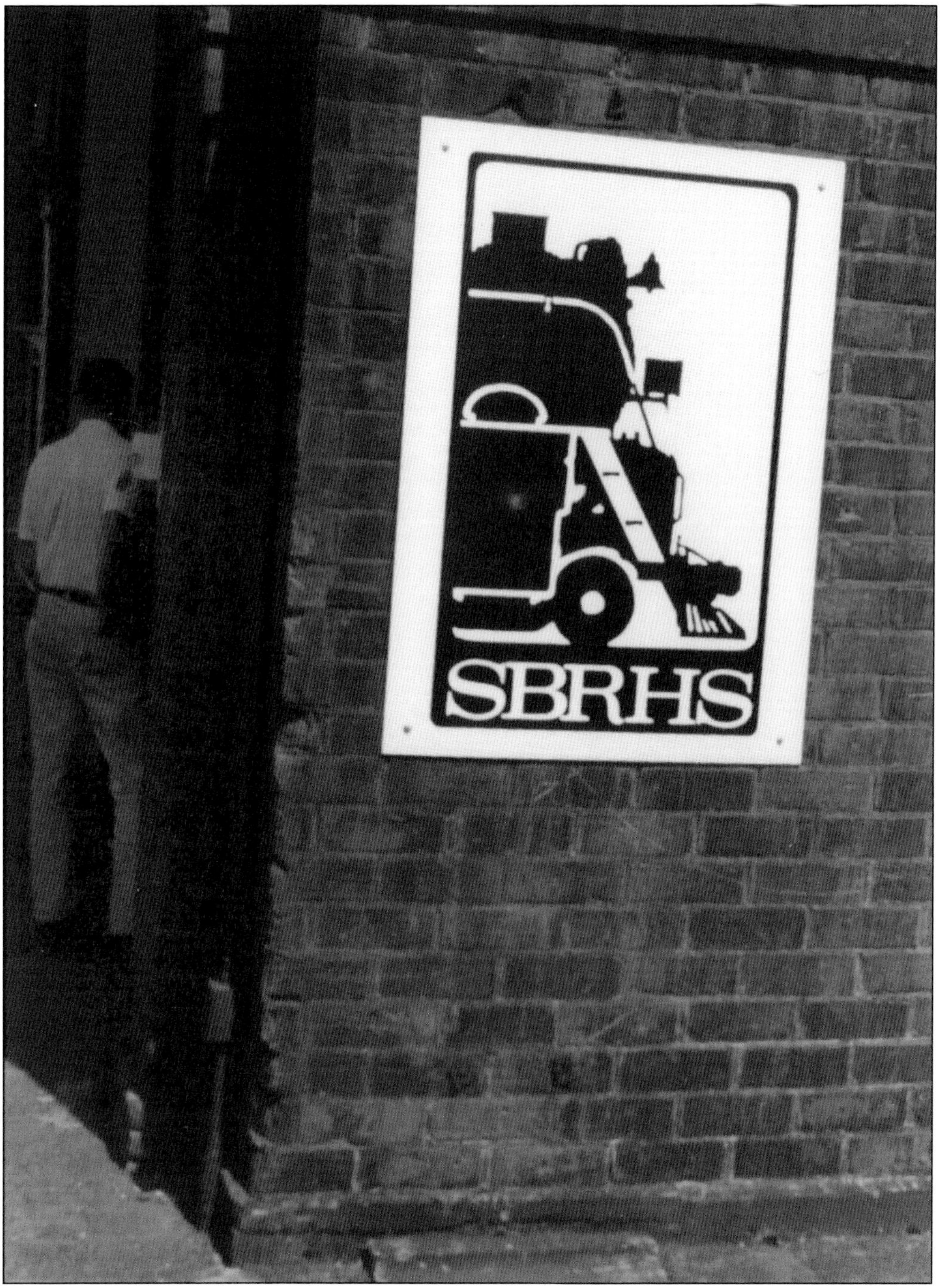

The San Bernardino Railroad Historical Society was formed in 1981 to restore locomotive No. 3751, a Fontana icon. The historical society's logo is seen on the side of the California Steel building where the engine was restored. This was the group's primary focus, with a secondary one of collecting historically relevant pieces of Santa Fe heritage for preservation. The engine, tool car, and reefer car have been part of the restoration. A primary task is protecting the engine from the elements. (Courtesy of the author.)

Seven

The San Bernardino Railroad Historical Society

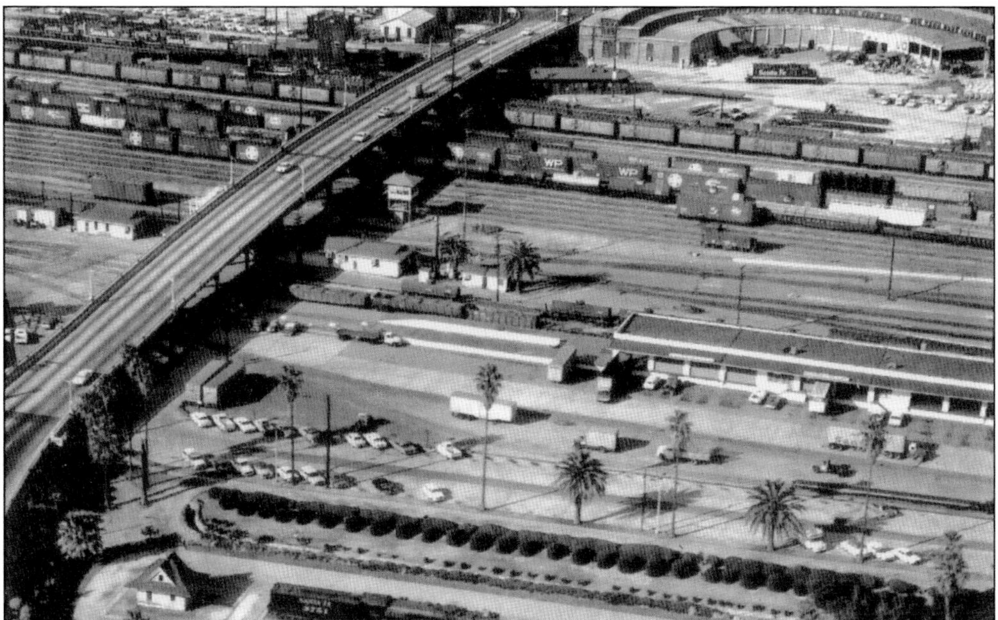

This aerial view of Viaduct Park includes Santa Fe locomotive No. 3751, which has become a Fontana icon. The San Bernardino Railroad Historical Society is composed of hundreds of individuals interested in the heritage and preservation of the "Steam Era."

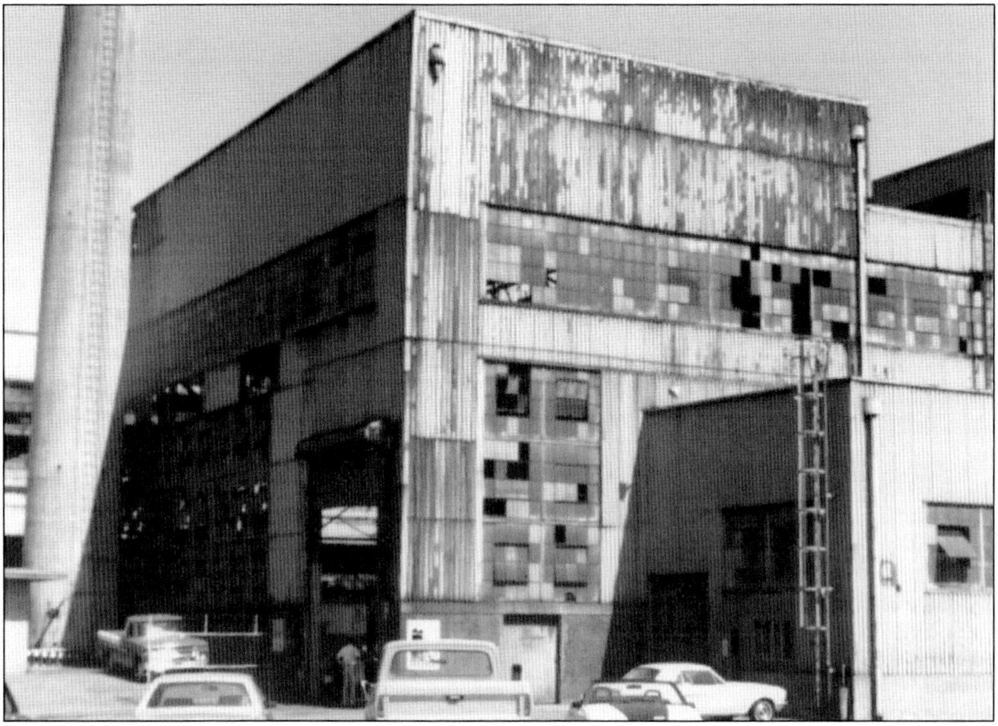

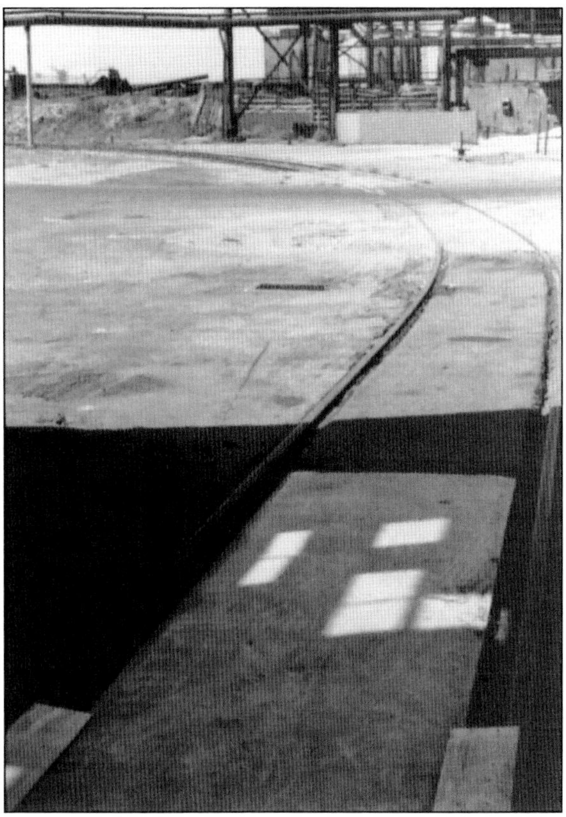

The Kaiser Steel Building, now owned by California Steel, was where the engine was restored. On April 30, 1986, No. 3751 was moved out of Viaduct Park and to California Steel Industries in Fontana.

The old track ran into the building. (Courtesy of the author.)

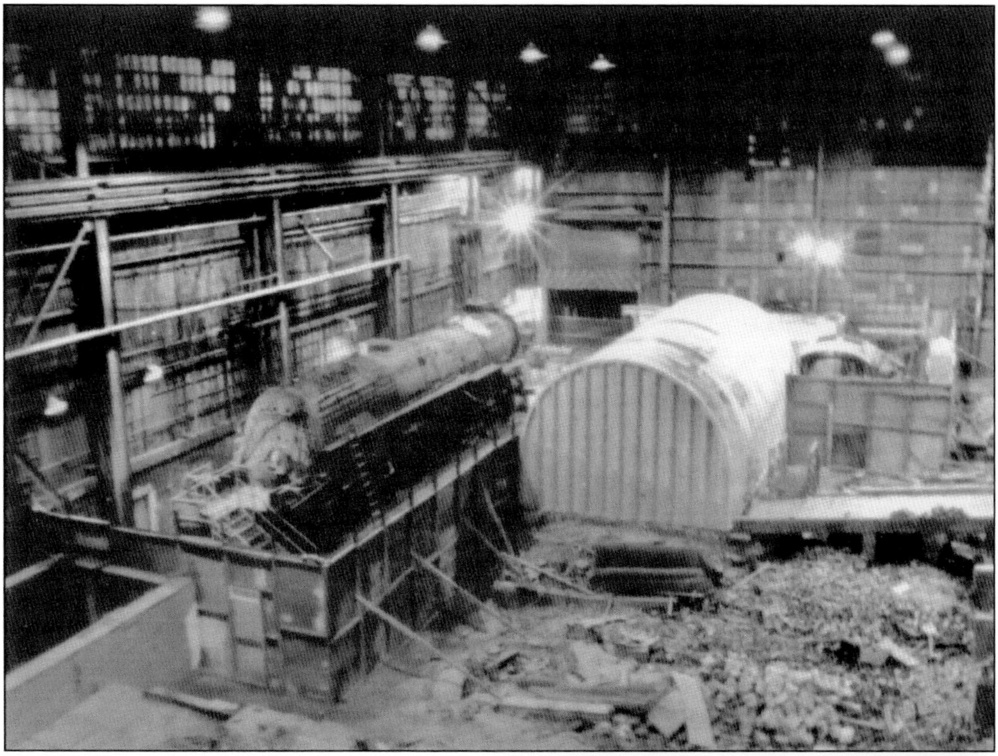

One small corner of CSI's disused, cavernous mill building was cleared to make room for Engine No. 3751's restoration. CSI donated use of the building and many services, making the engine's restoration possible. (Photograph by Dan Echeto; courtesy of the San Bernardino Railroad Historical Society.)

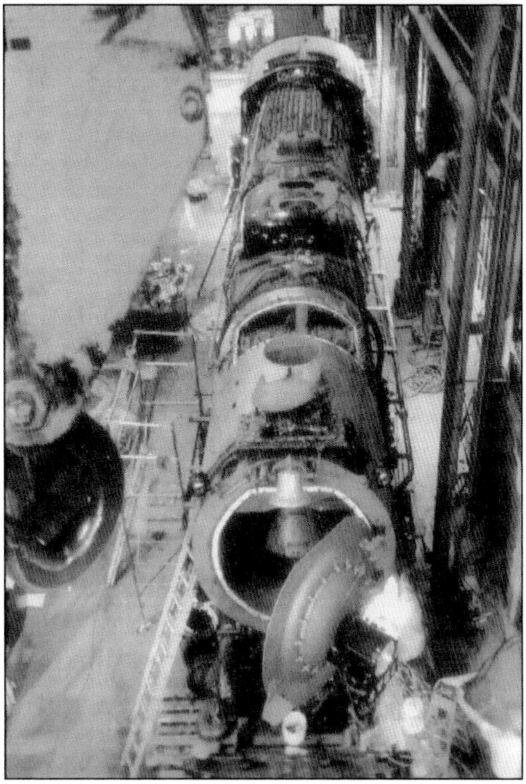

When this photograph was taken in February 1990, restoration of No. 3751 was well along its way, assisted in no small degree by the 15-ton traveling crane in the CSI mill facility. (Photograph by Dan Echeto; courtesy of the San Bernardino Railroad Historical Society.)

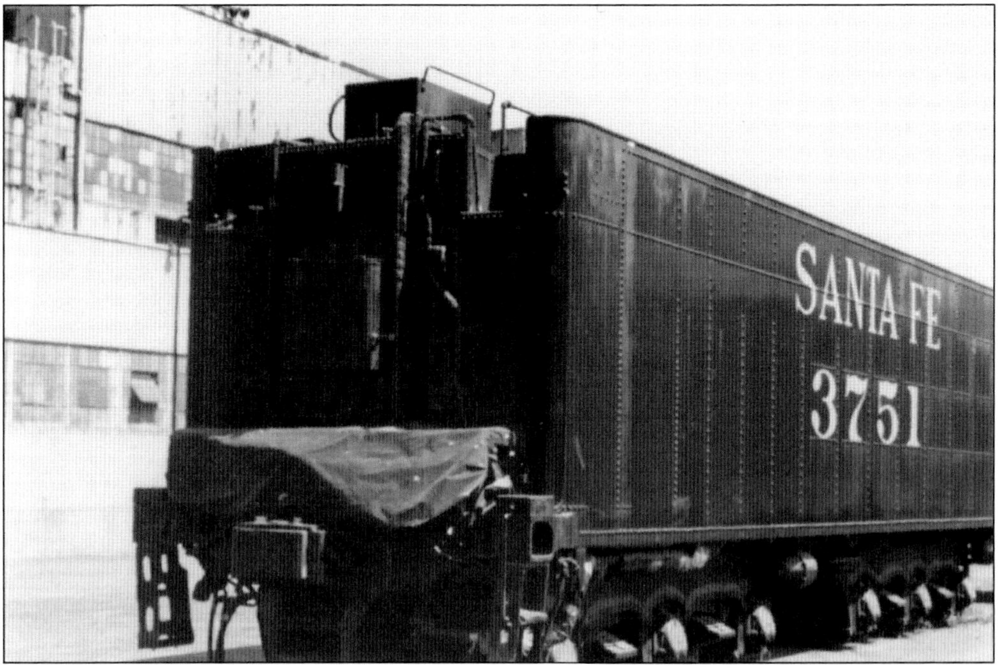

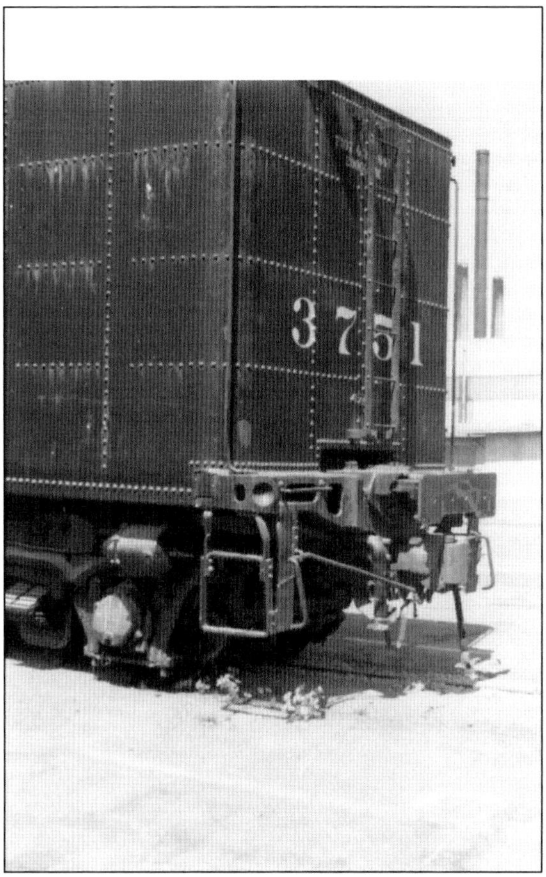

This is the timeline of No. 3751: In 1926, Santa Fe Railway, in cooperation with Baldwin Locomotive Works, designed Santa Fe's first 4-8-4 model as a coal burner; in May 1927, the 3751 emerges from the erecting hall as the Santa Fe and Baldwin's first 4-8-4; in 1936, No. 3751 is converted to burn oil at the San Bernardino Shops; in 1938, No. 3751 receives a larger roller-bearing-equipped tender with capacities of 20,000 gallons of water, and 7,107 gallons of oil and was designated as a standby for the Super Chief; On May 7, 1939, No. 3751 hauls the first named passenger train, *The Scout*, into the new Los Angeles Union Passenger Terminal on opening day. The car at left was stored outside during restoration of the engine. (Both photographs courtesy of the author.)

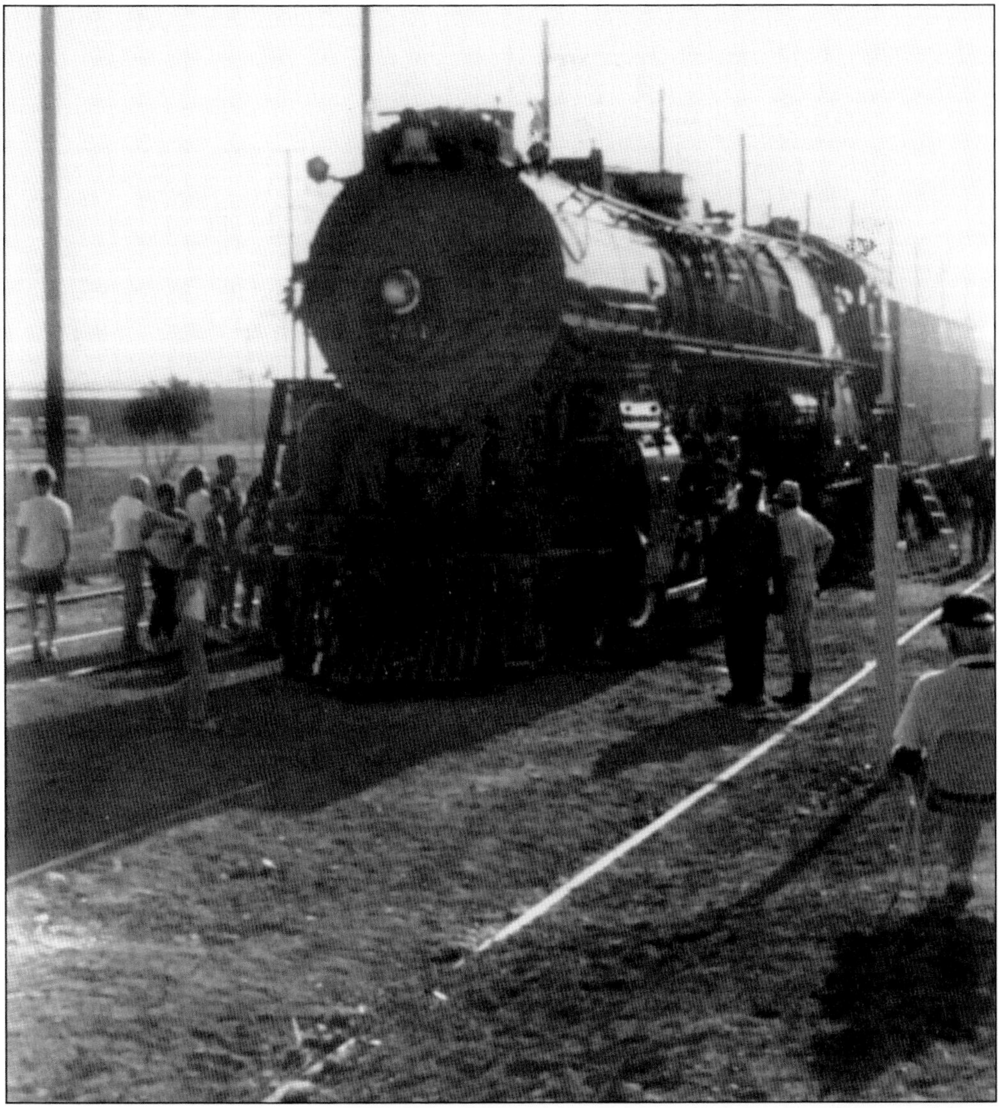

In 1941, No. 3751 is further modernized for high-speed operation with a new frame, an 80-inch drive, and roller bearings (its highest recorded speed was 103 miles per hour); on August 23, 1953, the locomotive pulls the last steam-powered passenger train from San Diego on August 23, the engine's last revenue run before being stored serviceable temporally at Los Angeles's Redondo Junction Round House; in 1957, No. 3751 is officially retired from the Santa Fe locomotive roster and donated to the City of San Bernardino; on May 14, 1958, No. 3751 is placed in Viaduct Park, San Bernardino, on a static display, becoming the oldest existing 4-8-4 as the first 12 that were built were scrapped by the Northern Pacific; in 1982, The San Bernardino Railroad Historical Society is formed with the intention of protecting the locomotive and possibly restoring her to operational condition; in 1985, No. 3751 is sold to the San Bernardino Railroad Historical Society by the City of San Bernardino for $1 with conditions that the society restore and operate the locomotive; on April 30, 1986, the locomotive moved to CSI for restoration; on August 15, 1991, No. 3751 returns to steam operation and pulled onto a siding at CSI along San Bernardino and Cherry Avenues, stopping traffic and drawing thousands of onlookers. Television helicopters arrived almost immediately. (Courtesy of the author.)

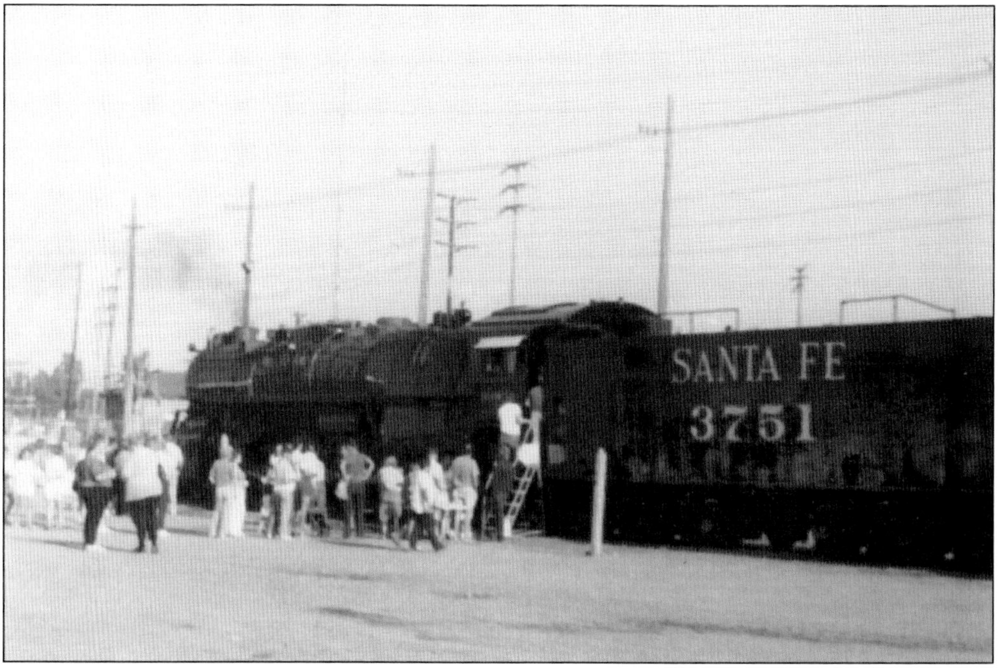

The railroad historical society gave a tour of the cab of the engine to anyone willing to climb the ladder. (Courtesy of the author.)

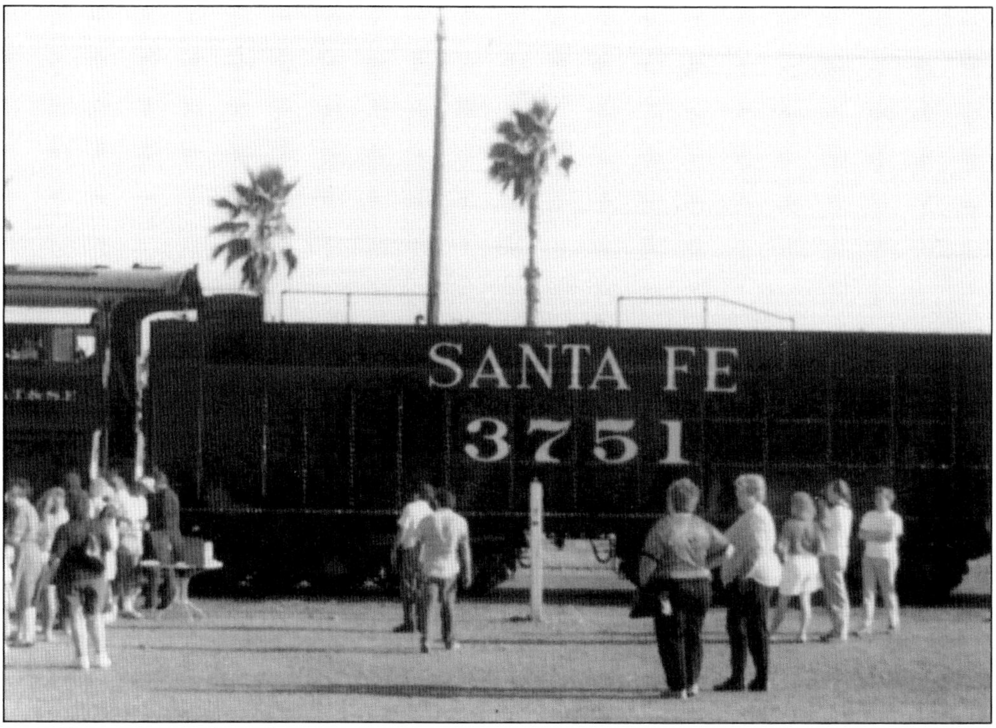

Rae Ellen Anicic and Sandy Rezes stand under the "1" of the No. 3751 on the car. This photograph was taken during a trial run of the engine to see if everything was in working order—which it was. (Courtesy of the author.)

Eight

California Speedway

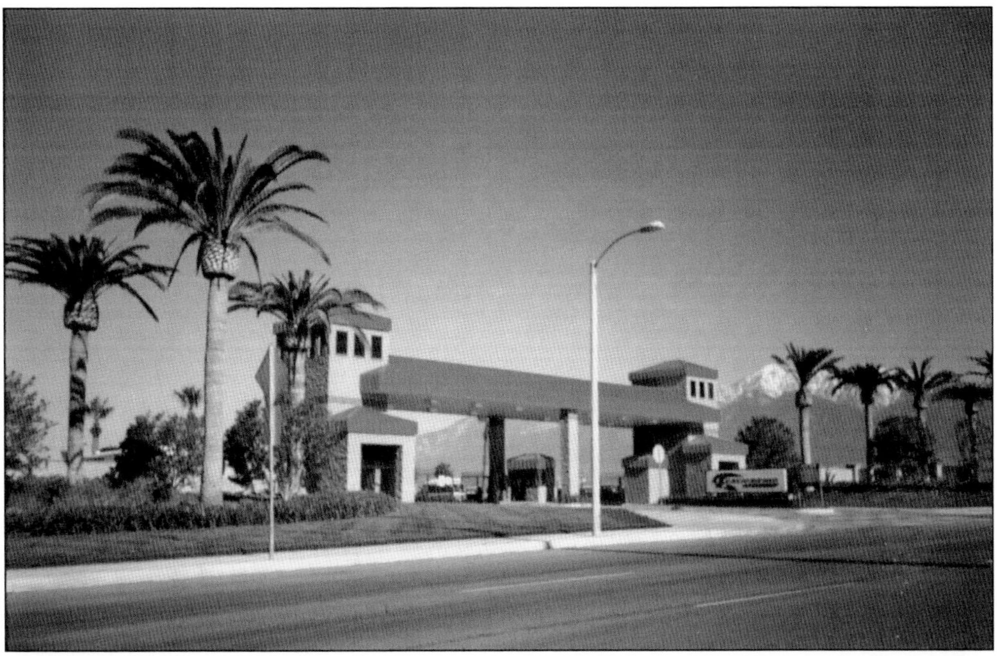

The main entrance to the California Speedway with its old logo is seen with the snow-capped San Gabriel Mountains in the background. The first five months of construction on the California Speedway consisted of dismantling the old Kaiser steel mill. More than one million tons of concrete and rubble was removed, and thousands of tons of scrap metal were hauled to scrap yards. Two million cubic yards of earth were moved to shape the two-mile, tri-oval track since construction began. Approximately 82,000 linear feet of storm, water, and sewer piping was installed. (Courtesy of the author.)

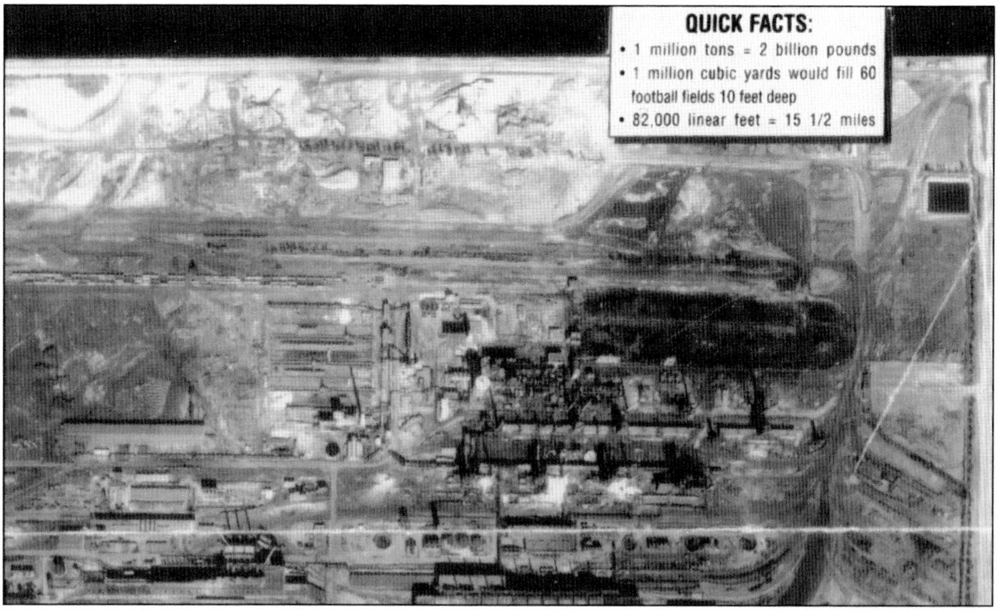

This aerial of Kaiser Steel was taken before construction began in 1985. On November 1, 1993, Kaiser Ventures, Inc., and Penske Speedway, Inc., agreed to prepare the site according to state environmental laws. On April 20, 1994, the official announcement of the Penske/Kaiser joint venture to construct a permanent super-speedway in Southern California was released. On April 26, 1995, the San Bernardino County Board of Supervisors unanimously approved the project, and on November 22, 1995, construction began. On April 1, 1996, the Kaiser steel-mill demolition was complete. (From Speedway information sheet.)

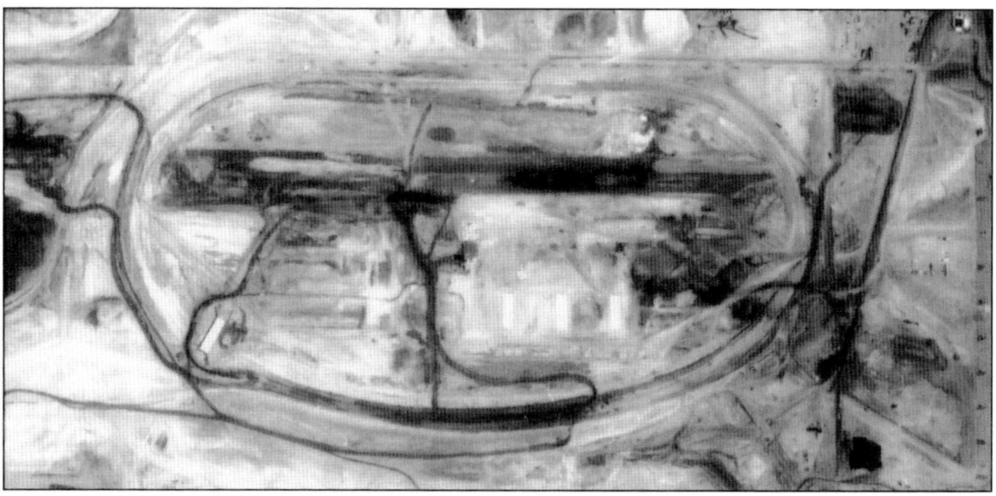

This aerial photograph depicts California Speedway on June 3, 1996. Foundations had been completed for the three garage buildings and the maintenance structure. Excavation was in progress for the foundation and underground piping for the 71 infield suites that run parallel to pit row. The grandstand permit was in place, and the steel structure for the grandstands was in fabrication. The tunnel, which extends under the track at the end of turn four, had been completed. The footing for the exterior crash wall had been completed from the top of the embankment of turn two, down the backstretch, and into turn three. The wall was constructed at a rate of 200 linear feet per day. (From Speedway information sheet.)

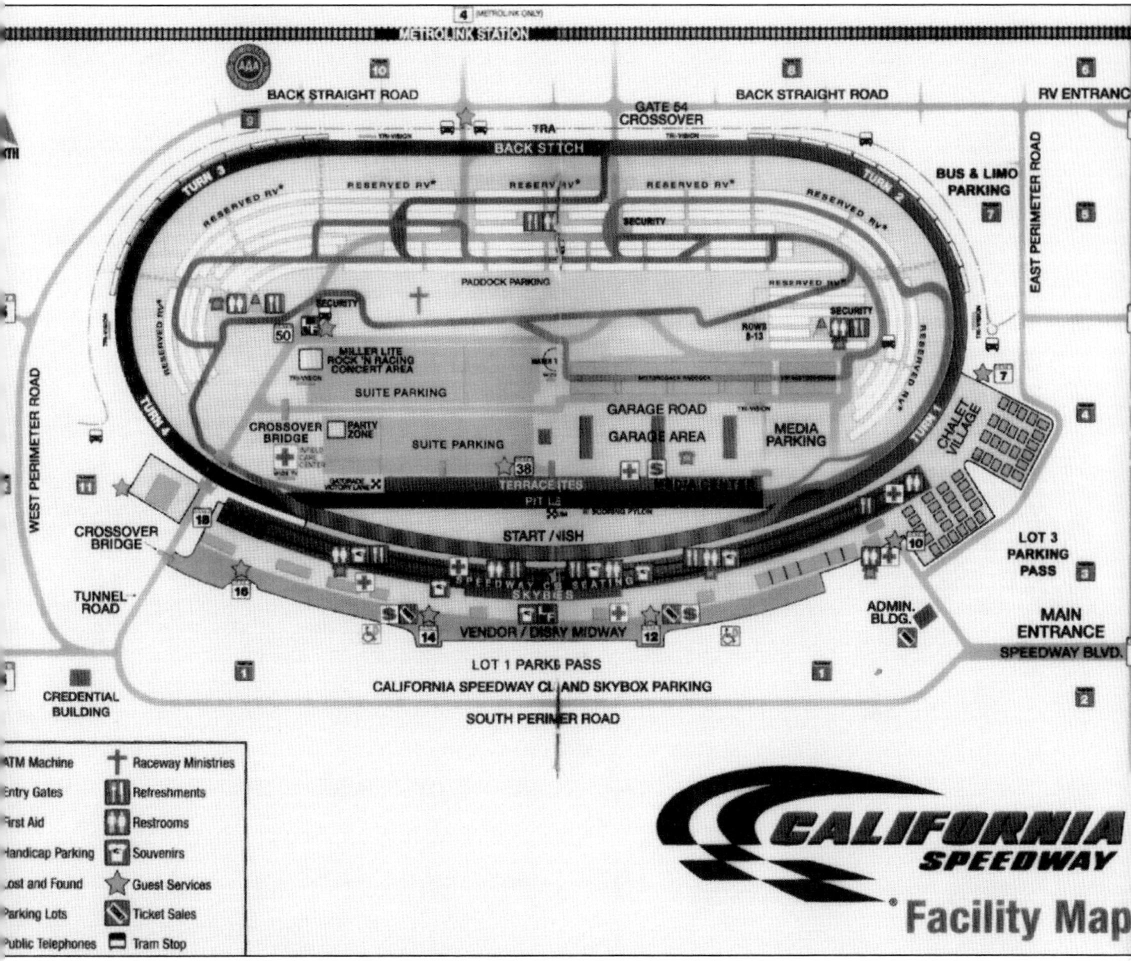

This is the California Speedway facility map with north at the top. On November 1, 1996, Penske Motorsports, Inc., expanded California Speedway from 475 acres to 529 acres with the acquisition of the Speedway business park frontage that is adjacent to the property. On January 10, 1997, Marlboro Team Penske driver Paul Tracy became the first driver to run on California Speedway's new two-mile oval race track. On June 20, 1997, the ribbon-cutting ceremony commemorated the official opening of the California Speedway. On December 8 1997, expansion began for reserved grandstand seating along the front stretch. An additional 15,777 seats were added to bring the total to 86,439 reserve seats.

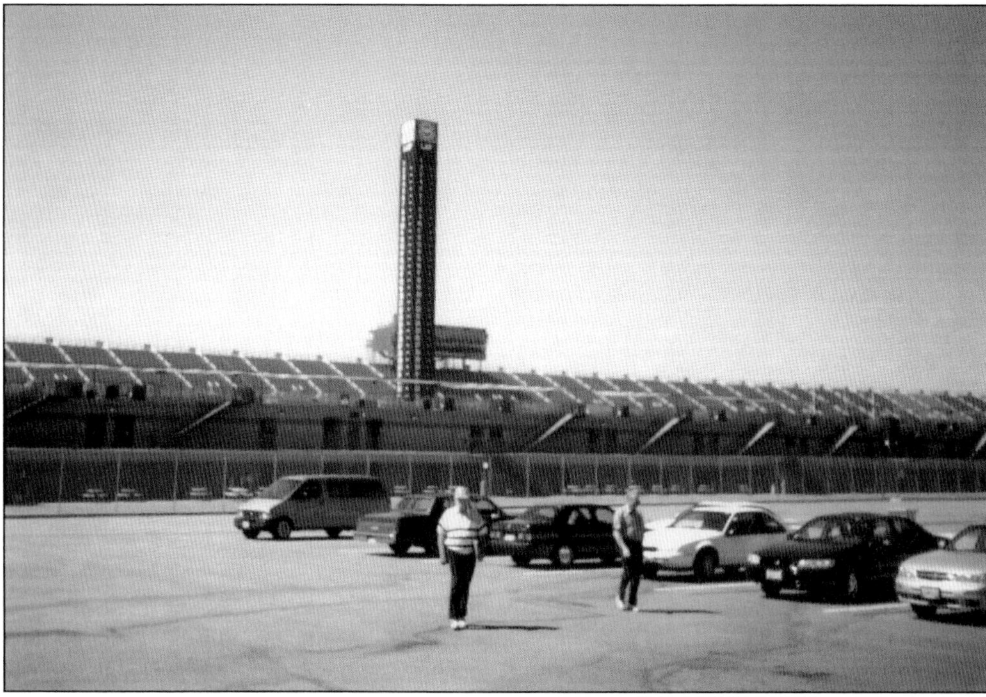

The grandstand is pictured here without the pending sky boxes, looking south. On May 3, 1999, construction on an additional 28 sky boxes began—the day after the California 500. The sky boxes were constructed on top of the main grandstand area and are connected to an original structure that housed race control and a press box. On November 1, 2000, expansion began on an additional 5,875 grandstand seats, located near turn four. An additional elevator tower was constructed. Permanent seating capacity reached 92,000.

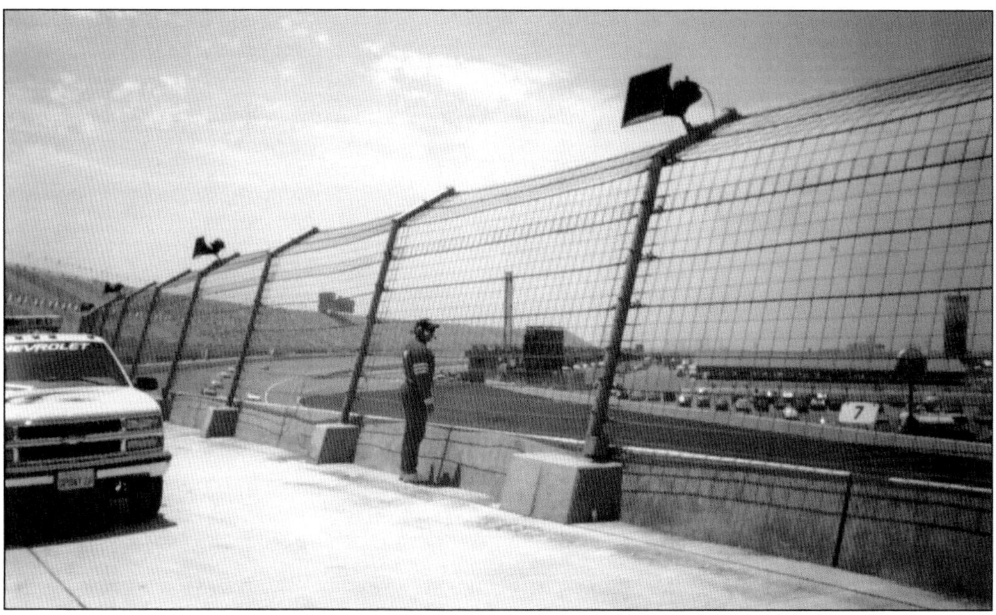

This is the fence at turn one, looking toward infield suites and the pit area.

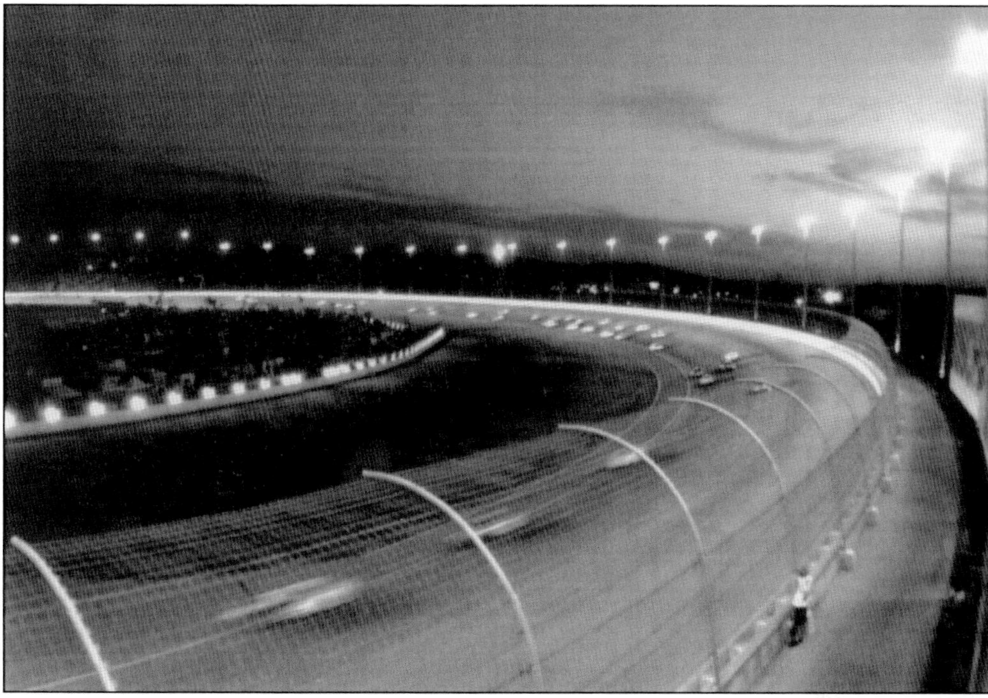

The California Speedway at night is a spectacular sight. On June 19, 2001, a quarter-mile street-legal drag strip was opened in Parking lot number 1, called California Dragway. A road course and motorcycle course were also opened. On April 24, 2003, the San Bernardino County Planning Commission approved a change in the speedway's conditional-use permit, allowing the installation of lights for night racing.

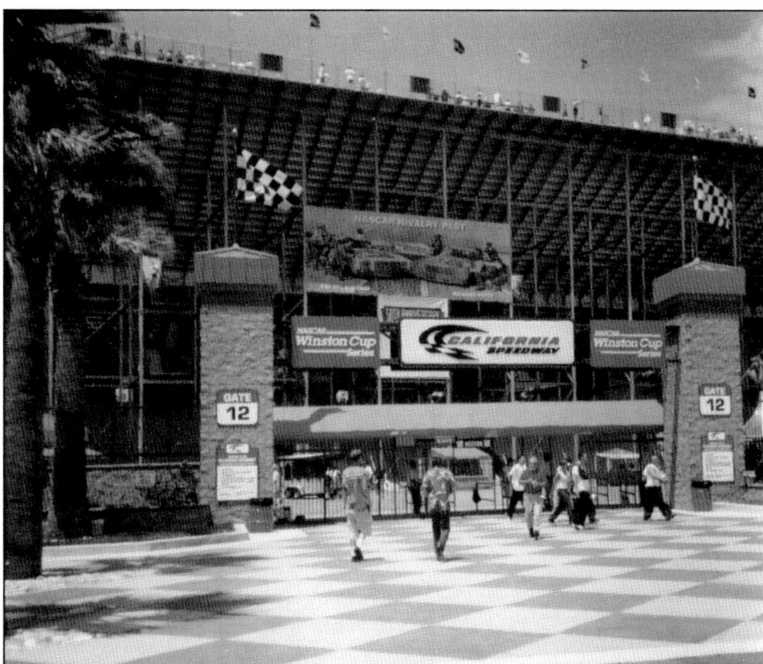

The south entrance to the grandstand area is seen here. The speedway scheduled races in 2004 to finish under the lights at night. California Speedway was the first world-class speedway to have lights installed. On October 13, 2003, construction of the Musco lighting system was initiated along the exterior of the two-mile on the backstretch.

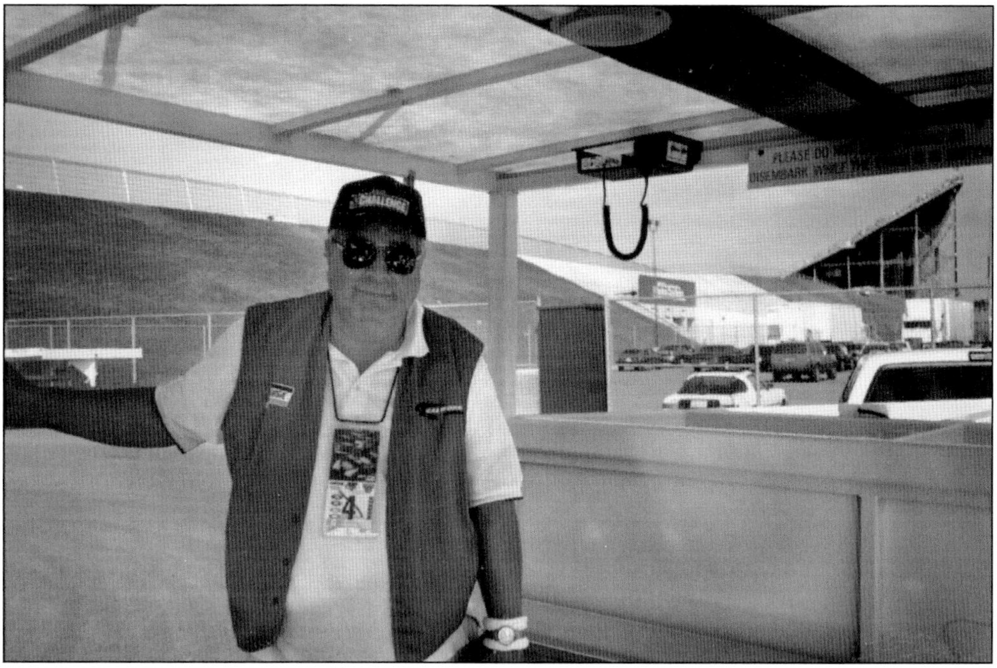

A free tram service was provided by California Speedway and manned by volunteers. In a mutual benefit arrangement, local groups provide volunteers during races, and the groups are then given donations according to the number of persons worked, reducing the need for permanent employees. Pictured here is author John Charles Anicic Jr., who worked at the speedway for several years for the Bloomington Preservation Foundation. (Courtesy of the author.)

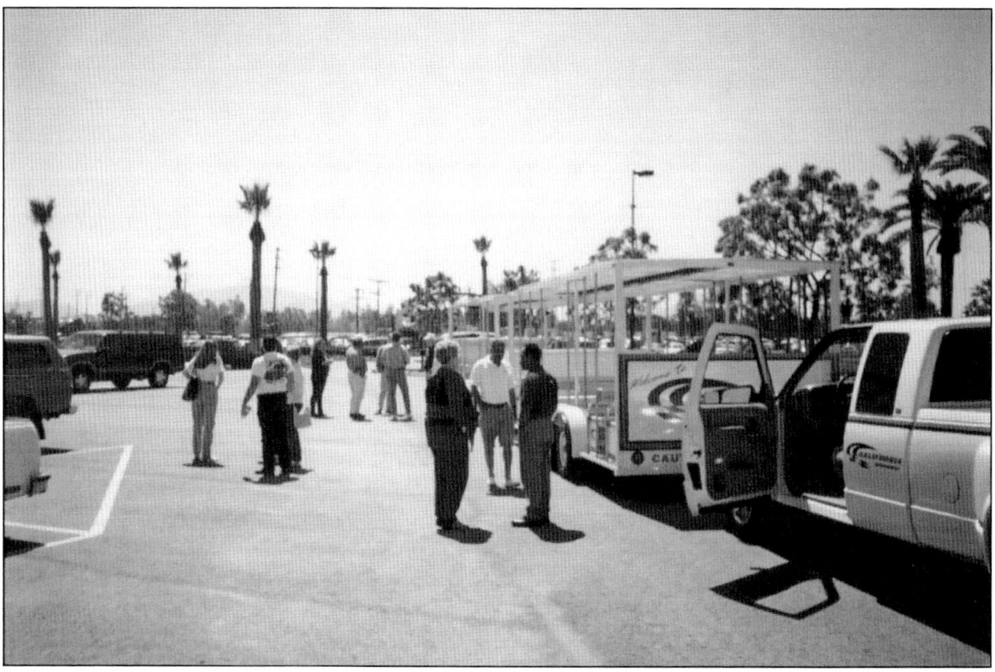

Loading of the trams is accomplished by volunteers. They also drive the trucks that belong to the speedway. Stops are located all around the area, even in the infield. (Courtesy of the author.)

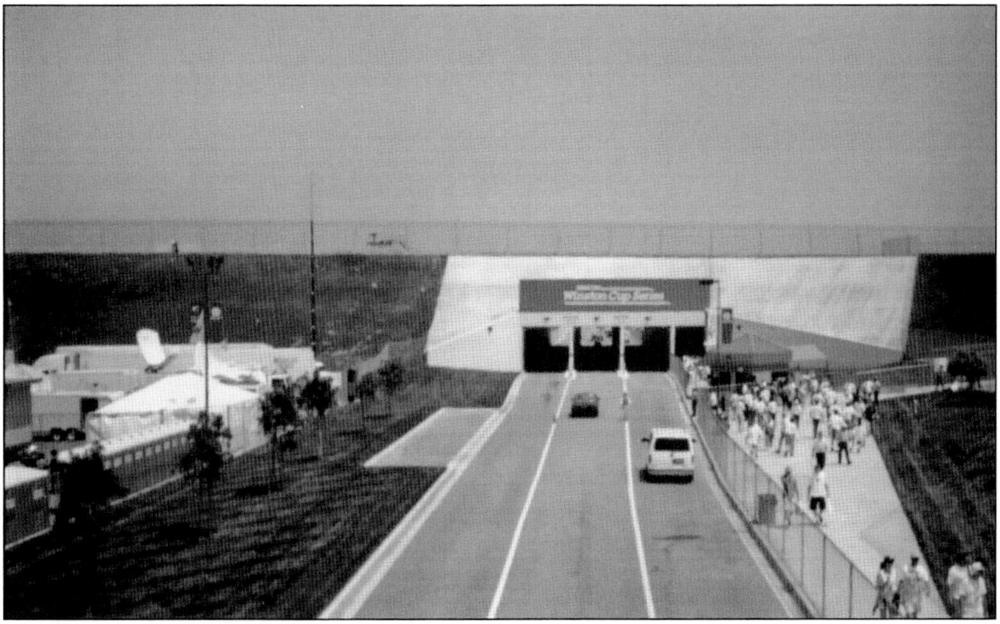

The tunnel to the infield and the recreational-vehicle area is pictured here during races. The garages, suites, and pits are also inside on the south end of the infield. (Courtesy of the author.)

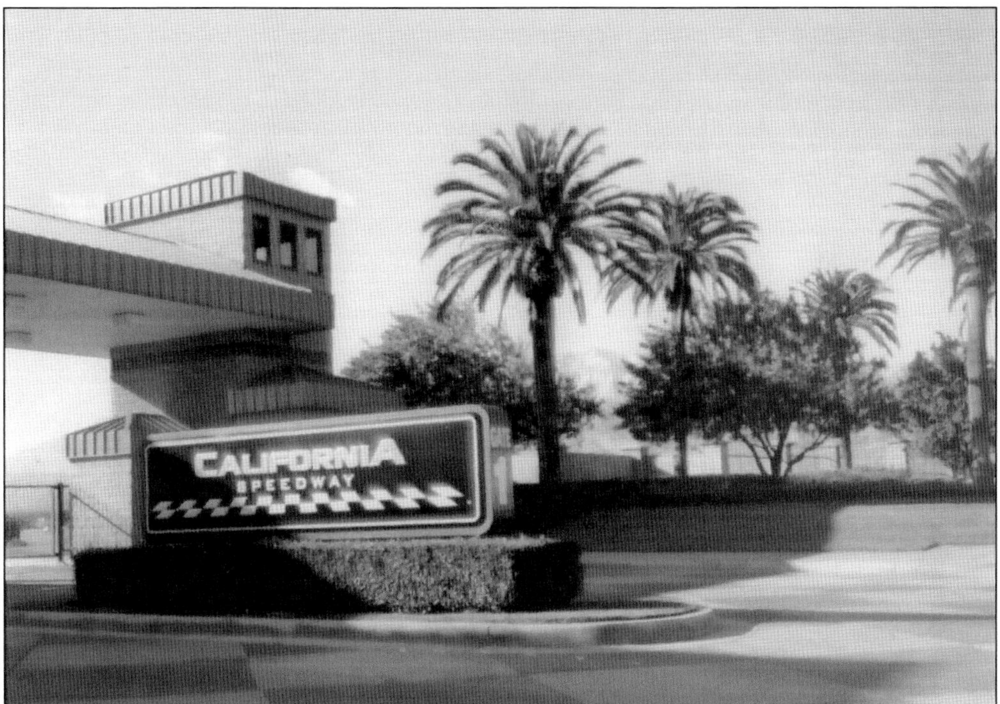

This is the new logo of the California Speedway at the main entrance on Cherry Avenue, looking northwest. (Courtesy of the author.)

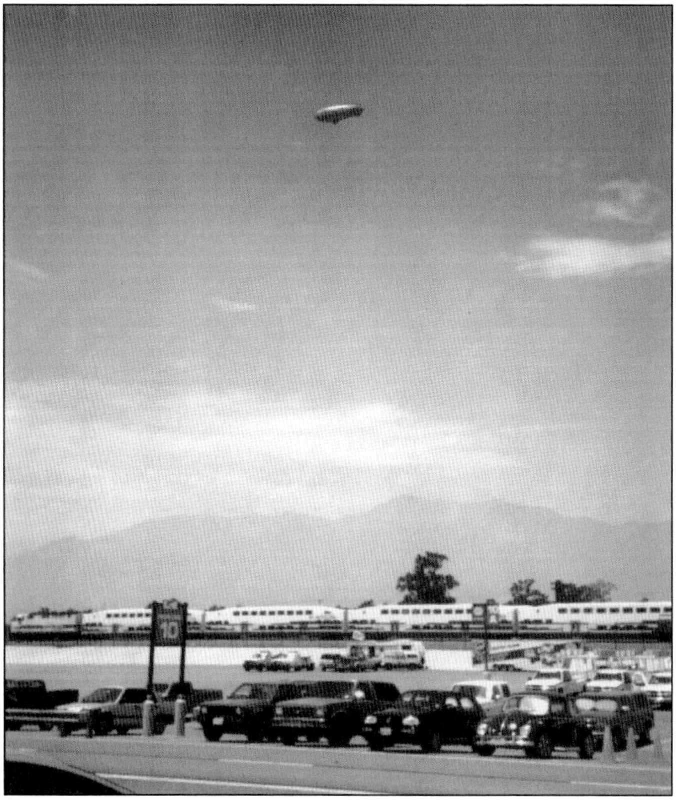

The Metrolink station is pictured here on the north side of the Speedway property. Trains come in from all over Southern California, bringing people to the races. The trains park next to the speedway until the races are concluded. Trams take the people to and from the trains. The Goodyear Blimp is over the speedway during many races, and other advertisers use small planes to pull large signs that advertise beer, cars, and other items. (Courtesy of the author.)

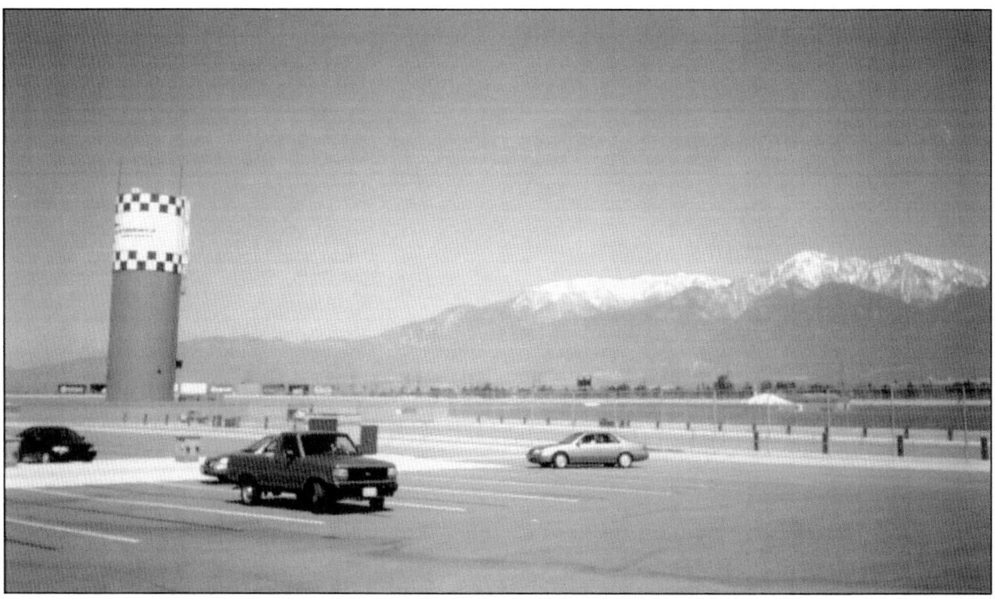

The water tower is the only landmark item that the speedway saved from the Kaiser Steel Mill heyday. The parking area here is covered with cars and RVs during races. The snow-covered mountains add to the beauty of the scene. (Courtesy of the author.)

AFTERWORD

There are some people I thought I would put back here at the end. They are mentioned here because of how special they are and what they mean to me.

Frank J. Videgar was a specialty man at Kaiser for a short time before going into the navy in 1953. He was given a leave of absence for four years, and then another four to go to college. He used his time well. He also plays the accordion, as did his family. They were great. He was also on the City of Fontana's federal grant-money committee for many years.

Russell Rogers worked at Kaiser for 30 years as an electrician in the rolling mills. He is the father of Rene, who I used to work with years ago. She's gone now, but I think of her often when I see her father.

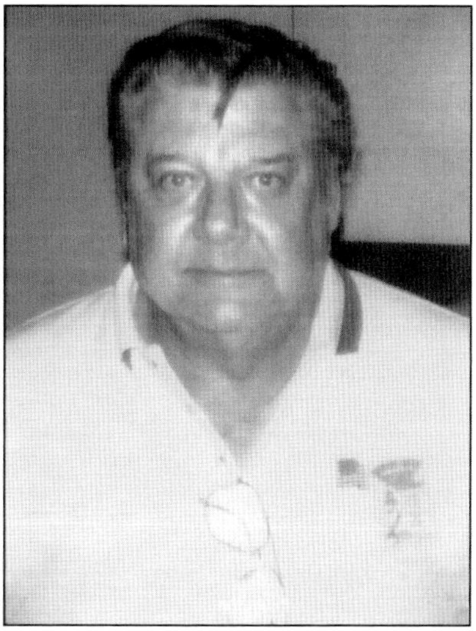

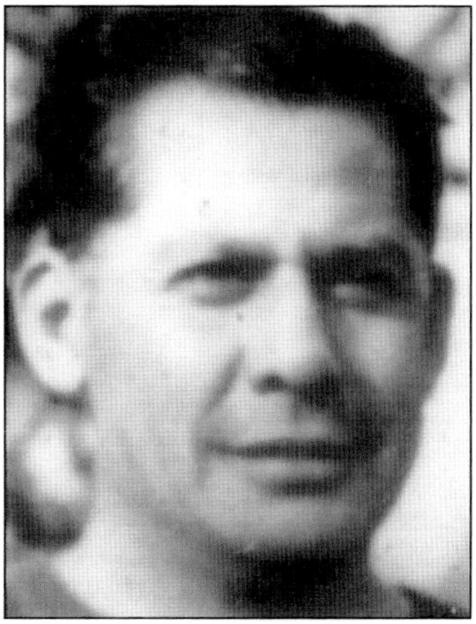

Alfred Flores worked at Kaiser Steel for 23 years as a heater helper. His wife recently passed away, and Alfred has been gone for some time now. Their five children now have both parents gone and need a lift up. I hope this will help. Good luck.

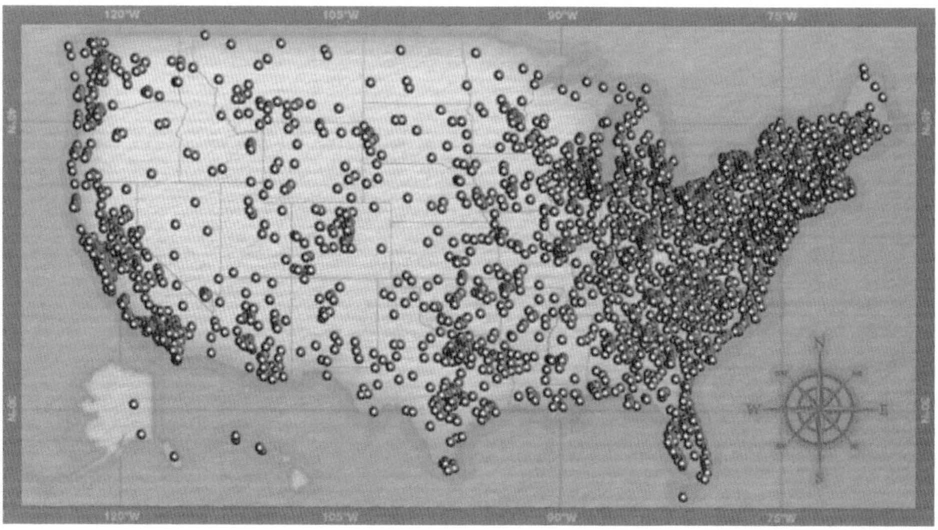